BAYESIAN METHODS IN STATISTICS

Sara Miller McCune founded SAGE Publishing in 1965 to support the dissemination of usable knowledge and educate a global community. SAGE publishes more than 1000 journals and over 800 new books each year, spanning a wide range of subject areas. Our growing selection of library products includes archives, data, case studies and video. SAGE remains majority owned by our founder and after her lifetime will become owned by a charitable trust that secures the company's continued independence.

Los Angeles | London | New Delhi | Singapore | Washington DC | Melbourne

Mel Slater

BAYESIAN METHODS IN STATISTICS

From Concepts to Practice

Los Angeles | London | New Delhi
Singapore | Washington DC | Melbourne

Los Angeles | London | New Delhi
Singapore | Washington DC | Melbourne

SAGE Publications Ltd
1 Oliver's Yard
55 City Road
London EC1Y 1SP

SAGE Publications Inc.
2455 Teller Road
Thousand Oaks, California 91320

SAGE Publications India Pvt Ltd
B 1/I 1 Mohan Cooperative Industrial Area
Mathura Road
New Delhi 110 044

SAGE Publications Asia-Pacific Pte Ltd
3 Church Street
#10-04 Samsung Hub
Singapore 049483

Editor: Jai Seaman
Assistant editor: Charlotte Bush
Production editor: Ian Antcliff
Copyeditor: QuADS Prepress Pvt Ltd
Proofreader: Neville Hankins
Indexer: David Rudeforth
Marketing manager: Ben Griffin-Sherwood
Cover design: Shaun Mercier
Typeset by: C&M Digitals (P) Ltd, Chennai, India
Printed in the UK

Library of Congress Control Number: 2021937241

British Library Cataloguing in Publication data

A catalogue record for this book is available from the British Library

ISBN 978-1-5297-6861-9
ISBN 978-1-5297-6860-2 (pbk)

To Mavi and Olivia.
To the memory of Laurence.

CONTENTS

LIST OF FIGURES

LIST OF TABLES

DISCOVER THE ONLINE RESOURCES

Bayesian Methods in Statistics is accompanied by a wealth of online resources to support your statistics teaching. Find them at: https://study.sagepub.com/slater.

For instructors:

- **PowerPoint decks** can be customised for use in your own lectures and presentations.
- **Screencast videos** demo step-by-step how to use R and Stan via kaggle.com so that your students can gain confidence navigating the platforms.
- Weblinks to all the **datasets** used the book enable your students to practice new statistical techniques at their own pace.

All the examples in this book and the corresponding data are available at Kaggle – see pp. 25–26 for details.

ABOUT THE AUTHOR

Mel Slater is a Distinguished Investigator at the University of Barcelona, and co-director of the Event Lab (Experimental Virtual Environments for Neuroscience and Technology). He was previously Professor of Virtual Environments at University College London in the Department of Computer Science. He was awarded the 2005 IEEE Virtual Reality Career Award: 'In Recognition of Pioneering Achievements in Theory and Applications of Virtual Reality'. He is Field Editor of *Frontiers in Virtual Reality*, and Chief Editor of the Human Behaviour in Virtual Reality section. He was awarded the A. v. Humboldt – J. C. Mutis Research Prize from Germany in 2020. He is a Fellow of the Royal Statistical Society.

ACKNOWLEDGEMENTS

My long-standing friend Professor Richard Wiggins (University College London) is a series editor of *The Sage Quantitative Research Kit* (due for publication during 2021).[1] About 3 years ago, he asked me to contribute a book on Bayesian statistics for that series. Although this book is not part of the series, it was inspired by it and might be thought of as something to read in conjunction with some of its volumes. I would like to thank Professor Wiggins for being such a constant friend over many years and for putting me on the path to writing this book.

I would like to thank the many people in my research group over the years who have carried out our experimental studies in virtual reality, which provided the inspiration for my data analyses. The people involved are too numerous to name. Similarly I would like to acknowledge the thousands of participants in our experimental studies over the past three decades.

The main software used for the preparation of this book has been Stan (https://mc-stan.org/) and R (https://www.r-project.org/), Rstudio (https://rstudio.com), MATLAB® (www.mathworks.com) for the simulation of data, and Stata (www.stata.com) for the production of some of the graphs. Adobe Illustrator and Photoshop were also used for formatting or joining separate graphs together in one plate. Microsoft® Word for Mac was used for the text, and Endnote (https://endnote.com) for the references.

I would like to thank Mavi for putting up with my staring at spreadsheets, R, and Stan code over many years.

Finally, I would like to thank the government and people of Borgonia for their cooperation throughout the writing of this book.

Mel Slater

[1]https://uk.sagepub.com/en-gb/eur/the-sage-quantitative-research-kit/book277911

PREFACE

Introduction to the Book

I started learning statistics during my undergraduate and master's degrees at the London School of Economics and Political Science (LSE). An outstanding teacher who comes to mind is the late Professor Alan Stuart, who taught me probability and distribution theory, although there was hardly any mention of Bayes during those times at LSE.

I joined the faculty of Queen Mary and Westfield College (Queen Mary, University of London) and was responsible for teaching courses such as linear models and advanced statistics in the Department of Computer Science and Statistics. During this time, I was introduced to the Royal Statistical Society GLIM Working Party by the late Professor Mike Clarke. GLIM (Generalized Linear Interactive Modelling)[2] was a software package for generalized linear models. Professor Clarke suggested that I become responsible for a computer graphics module of a future version of GLIM. This led me eventually to go into full-time teaching and research in computer graphics, and I stayed in Computer Science when the statisticians moved to the Department of Mathematics.

In the early 1990s, I had my first experience of virtual reality, and I wanted to understand why I had such as strong sense of being in the virtual world depicted through the head-mounted display, as if I were really there, even though the level of graphical realism was extremely low. I was also shocked by the fact that when I looked down towards myself in the virtual reality, I was invisible. Given my original training in statistics, I turned to experimental studies with people in order to begin to understand how virtual reality works at the perceptual and cognitive levels. My group at that time included Dr Martin Usoh and a bit later (now Professor) Anthony Steed, originally at Queen Mary and from 1995 at University College London. We embarked on a long journey of virtual reality research, and I'm still on that journey today, as is Professor Steed. In 2006, I moved to Barcelona as an ICREA Research Professor (www.icrea.cat) first at the Universitat Politècnica de Catalunya and then from 2008 at the Universitat

[2] https://en.wikipedia.org/wiki/GLIM_(software)

de Barcelona in the Faculty of Psychology, co-founding the Event Lab (Experimental Virtual Environments for Neuroscience and Technology, www.event-lab.org) with the neuroscientist and ICREA Research Professor Mavi Sanchez-Vives. For the analysis of the results of experiments, I used classical statistics (regression, analysis of variance, *t* tests, etc.) but became more and more disillusioned with this approach. There was mounting criticism in the literature of 'null hypothesis significance testing', but alternatives proposed such as the 'new statistics' (Cumming, 2014) didn't seem to me to solve the problems. Moreover, most of our experiments involved more than one response variable and therefore multiple statistical tests. As soon as more than one test is carried out, there is a loss of control of significance. I could not get past the paradox that if I carry out a significance test then all the other tests that I have done on the same experiment have to be taken into consideration when computing the true significance level. But why not the tests I did last week? Why not everyone else's tests? If we take the problem of multiple testing to its logical conclusion, then no test ever could be considered as 'significant'. The Bayesian approach seemed superior overall, so that in the last few years almost all of our research has employed Bayesian statistical analysis for the analysis of results of experimental studies.

The intention of this book is to introduce probability and statistics from a Bayesian point of view. The aim is to make it a practical book that will teach readers how to carry out Bayesian analysis but at the same time understand the fundamental concepts without going deep into the mathematics. It begins with an introduction to probability theory through the axioms of probability. The axioms provide a framework for rigorous calculations of probabilities, but they do not at all indicate how to ever assign a probability. The Bayesian interpretation is that probabilities are initially assigned subjectively, providing a quantitative indication of the assignee's degree of belief or prior knowledge about the event or proposition in question. Bayes' theorem is a simple consequence of the mathematical formulation of probability and yet provides the foundation of statistical inference. After Bayes' theorem is introduced, the remainder of the book illustrates how it is used in a variety of different situations. This is like the famous Golden Rule of ethical behaviour: 'What is hateful to you, do not do to your fellow: this is the whole Law; the rest is commentary. Go and learn!' (Hillel). So we can say the same about Bayes' theorem: there is Bayes' theorem at the heart of statistical inference, and all the rest is commentary, go and learn.

Bayesian statistics is only practically possible through computer simulation to obtain solutions to complicated problems of integration. The simulations are obtained through Markov chain Monte Carlo methods, and these are introduced in the final chapter. That chapter also compares various aspects of the Bayesian and classical approaches to statistics, an interpretation where the probability of an event is considered exclusively as the long-run frequency of its occurrence, and where, therefore, probabilities cannot be assigned at all to propositions or hypotheses.

Readers would benefit from a prior understanding of some basic mathematics, in particular mathematical functions – for example, the shape of the curve that this function might have: $f(x) = e^{-x^2}$; and the meaning of the integral $\int_a^b f(x)dx$. Some familiarity with the notation for sets would be useful, such as $\alpha \in A$ (α is a member of the set A), or $A \cup B$, or $A \cap B$ (union and intersection, respectively), although these are explained in the text. However, this is not a mathematical book, there are few derivations and where there are they can be skipped – they are included for those with an interest and who are able to follow them.

Since solutions to problems in Bayesian statistics typically require computer simulation to solve integrals, software must be used. In this book, we use the statistical modelling language Stan (https://mc-stan.org; Stan Development Team, 2011–2019). This is a programming language that supports description of the data to be analysed, specification of parameters, formulation of the model and predictions, and of course, solutions. Stan is accessed through another interface – here we use the R language and environment (R Core Team, 2013) and in particular the R interface to Stan (rstan): https://cran.rproject.org/web/packages/rstan/vignettes/rstan.html

R itself is not directly taught within this book; however, mostly very simple use is made of it, it should be straightforward to understand, and following the examples in this book should help the reader to also learn some basics of R.

The early chapters of the book rely solely on simulated data. In order to indicate this, such data are always ascribed to the mythical country of Borgonia. In the later chapters, some real data arising from virtual reality experiments have been used. The sources of the original papers where the experiments are described are cited.

Online Resources

All the examples in this book and the corresponding data are available online via the 'kaggle' system (www.kaggle.com). This allows readers to interactively execute R and Stan code step by step and change the commands to see what happens. The website addresses for the examples are given at the end of each chapter, but they are all of the form www.kaggle.com/melslater/slater-bayesian-statistics-*n*, where *n* is the chapter number (sometimes it might be *n*a and *n*b where there is more than one web page). For example, if you go to www.kaggle.com/melslater/slater-bayesian-statistics-3, you should see the top of the web page as follows:

The first thing you should do is register so that you have your own login. Then you can click the 'Copy and Edit' button at the top right corner and you should see a new web page:

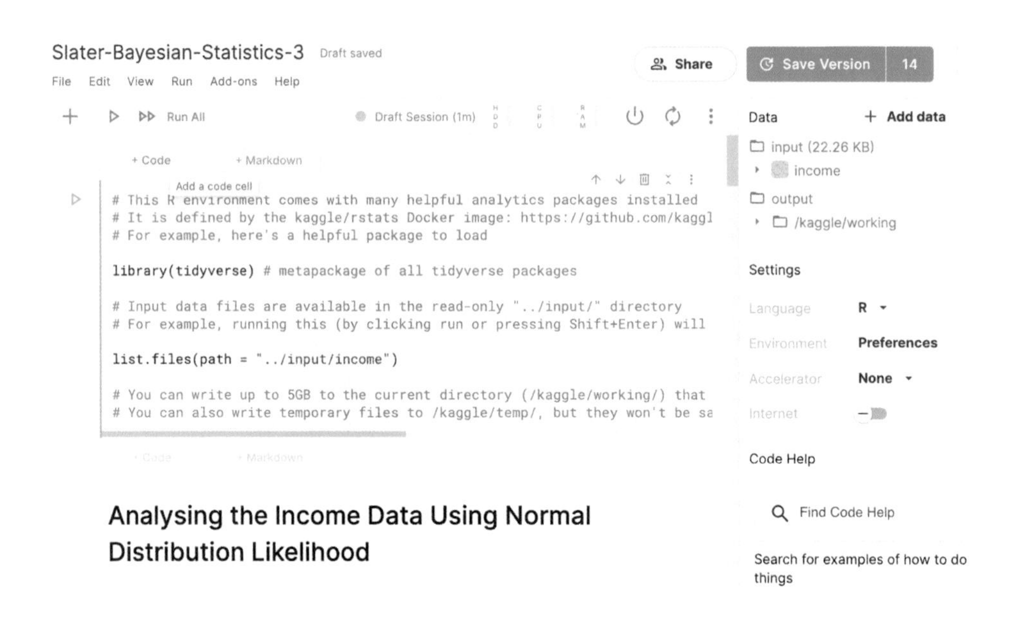

At the right-hand side you can see the name for the data file called 'income' that can be accessed. You will also see that the R language is already selected.

Each block of the page is executable. If you click on the little arrow to the left of the block, it will execute that block. If you want to add a new block, select '+Code' if it is to be a block only with code, or '+Markdown' if it is only for commentary. Readers are strongly advised to follow all examples in the book through this Kaggle interface.

The programs in the book are labelled as Box numbers – for example, Box 4.2 is in Chapter 4 and is a segment of code, and this can be found in the corresponding Kaggle for that chapter.

Who This Book Is for and How to Use It

Readers of this book will typically have had a basic course in statistics, so that there is familiarity with means, standard deviations, correlations, *t* tests, and so on, and will probably have been introduced to some statistical software. This book is aimed at helping you towards the direction of actually carrying out quite complex Bayesian analysis, while understanding the conceptual framework, but without going into the rigorous aspects that might be covered in a course on mathematical statistics. It is envisioned that the book will be useful for engineers and computer scientists who carry out experimental studies, for example in the domain of human–computer interaction including, of course, virtual reality. It will be also useful for researchers in the psychological and social sciences, assuming some basic mathematics as outlined above. In particular, the book is aimed at undergraduate final year students studying research methods, carrying out projects, or master's or PhD students in these fields, and generally for researchers who want to learn to understand the concepts of and carry out Bayesian statistical analysis.

The very early origins of the book (more than 15 years ago) were my notes for teaching a class on research methods in an advanced master's course at University College London that specialized in computer graphics and virtual environments in the Department of Computer Science. However, none of those original notes have survived into this book, which has been wholly written from scratch. This does illustrate, though, the type of audience to which this book is aimed.

Readers should read the book closely in conjunction with the online resources through Kaggle described above. When a program is shown in the book, find the corresponding one in Kaggle, and execute it to see what happens, and check the output. Be bold and change the Kaggle program, and see what happens if you change some of the parameters or their distributions. If you have your own data that are similar in design to the issue being discussed, then try using them. The more that you actually manipulate and explore the code, the more you will understand. If you don't know R, it doesn't matter – just execute each piece of R code to see what happens. Mostly very simple use is made of R: it is just used as a command interface to Stan. There are only two actual R programs (functions) in the book, in Chapters 1 and 7. These are not integral to understanding the surrounding concepts but are just illustrations. If there is some R expression that you don't understand, it is very simple to find out what it means. For example, suppose you come across the term 'list': just type 'R list' into your search engine (I use Google), and it will present many pages where lists are described.

R and Stan are presented online through Kaggle. Eventually, you might want to actually install R and Stan on your computer and use it that way. The Stan program in Kaggle will be identical, but the way to access files will be slightly different. You can find extensive notes and examples on the rstan web page https://mc-stan.org/users/interfaces/rstan

Overall, the message is that this should be thought of as an 'interactive book' – one where there is presentation of the basic ideas and examples, but the rest is up to you, through exploration and trying things out.

One

PROBABILITY

Introduction

Statistical science is concerned with understanding something about the whole by only knowing the part. Typically, the 'whole' is specified by a model or statement which is an abstraction about the world. For example, 'Adult humans aged 20 are taller than adults aged 16'. This is a statement about the world, which may or may not be true. But even this is difficult – since what does 'truth' mean here? Were there one 16-year-old taller than some 20-year-olds, would this make the statement false? This is probably not what the statement is supposed to mean. So we could be more precise and say 'On the whole adult humans aged 20 are taller than adults aged 16'. What does 'on the whole' mean? Now we have to be completely specific about what this expression means, otherwise there can be disagreements about whether the statement is valid or not. So we could say 'The mean height of adults aged 20 is greater than the mean height of adults aged 16', where 'the mean' has a precise definition: the mean $\bar{x}$ of a set of numbers x_i, $i = 1, 2, ..., n$, is the sum of those n numbers divided by n. Now our statement has a precise meaning (up to the point where the exact specification of how to measure a height is agreed – but we will assume for simplicity that this is agreed). However, there is still a problem, since to properly answer this question we would need to get the heights of every single member of the population aged 16 and 20. This has some tricky aspects since during the course of these extensive measurements, some people aged 16 when we started would be 17 by the time we measured them, and so on. In any case, what is meant by 'the population'? Everyone in the world? A specific country? A region? Our simple question seems to be very complicated to answer.

It is clear that, to approach this problem, we need something like a census time C: a period where we would consider all people aged 16 or 20 during that period. Moreover, we need a specific population target: for example, considering all cases in a particular country, say Borgonia. So now our statement is 'For the period C, the mean height of 20-year-olds in Borgonia is greater than the mean height of 16-year-olds in Borgonia'. This is now reasonably precise. But – there are a lot of 16- and 20-year-olds in Borgonia! Millions. We just do not have the resources to measure all N_{16} 16-year-olds and all N_{20} 20-year-olds. In fact, we only have resources and time to measure $n_{16} = 400$ 16-year-olds and $n_{20} = 400$ 20-year-olds. So we gather our team together, and during the period C we do all these measurements. We select the samples randomly from the Electoral Registers (16-year-olds have the vote in Borgonia), and indeed we find that $\bar{x}_{20} > \bar{x}_{16}$, where $\bar{x}$ is the usual notation for the mean of a sample. Recall, however, that this finding applies only to the tiny proportion of the whole population, so although our result is true for the sample, it still may not be true of the population as a whole. By bad luck, it could just be that our sample happened to have particularly tall 20-year-olds and particularly short 16-year-olds, and that this doesn't reflect the situation of the true population. This is where probability is important. We want to be able to say something like 'There is a high probability that $\mu_{20} > \mu_{16}$' based on our data. What is probability? What is 'high'? Here μ_{16} and μ_{20} are the population or 'true' means, whereas the observations are used to compute the sample means $\bar{x}_{16}$ and $\bar{x}_{20}$.

On to the discussion of probability.

Probability is a measure of the 'degree of belief' in the truth of some proposition, or the occurrence of some event. Probability theory provides rules that determine how to construct a probability of a more complex event or proposition from known probabilities of the constituent, simpler or elementary, events that make up the complex one.

A 'domain' is the set of all possible elementary events or propositions under consideration with respect to a given problem. We will use the word 'event' to cover 'proposition' as well. An 'elementary event' is one which, within the given problem area, cannot be broken down into yet more constituent events.

For example, in the tossing of two coins, it is possible that the coins might roll away and get lost, land on their rims, or be stolen by someone before the result ('Head' or 'Tail') of the toss was known. Nevertheless, the *domain* of the tossing of two coins would normally be the set $\{HH, HT, TH, TT\}$.

The 'elementary event' in the heights example is a bit more complex. The domain of a height is a range in the set of positive real numbers. It is difficult to precisely define this range, since whatever minimum we propose could be contradicted by some as yet unseen observation, and similarly, any maximum we propose could likewise be empirically contradicted. It is probably the case that no one has ever had a height of more than 15 metres (though according to Genesis 'there were giants in the earth in those days'). So we might take the domain as all positive values less than 15. A succinct way to write this is using the notation of sets:

$$\{h \mid 0 < h < 15\}$$

The '|' is interpreted as 'conditional on', or simply just 'for ...'. Sets are usually indicated by curly brackets {...} where the ... specifies the members or elements of the set. For example, $\{HH, HT, TH, TT\}$ is the set consisting of the elements that form the possible outcome of two tosses of a coin (or tossing two coins). Note that the order of these elements doesn't matter: for example,

$$\{HH, HT, TH, TT\} = \{HT, TH, HH, TT\}$$

or any other permutation of these four elements. So $\{h \mid 0 < h < 15\}$ means the set of all numbers (denoted by h) conditional on h having the property on the right-hand side of the '|', meaning that h has to be positive ($h > 0$) but less than 15 ($h < 15$). A simple way to read it is 'the set of all h for h between 0 and 15', or just 'all numbers between 0 and 15'. Of course, h can be replaced by any other symbol; we chose h because it can be thought of as a shorthand for *h*eight.

Alternatively, we might choose as the domain for this example simply the set of all positive real numbers $\{h \mid h > 0\}$. But actually, we will find later that using our prior knowledge, in this case that humans tend to have a maximum height, is advantageous in statistical inference (inferring about the whole from the part) compared to assuming all possible positive values (for heights in this case).

To specify that something is an element or member of the set, we use the notation $\in$. So clearly,

$$HH \in \{HH, HT, TH, TT\}$$

meaning that HH is a member of the set $\{HH, HT, TH, TT\}$.

But $H \notin \{HH, HT, TH, TT\}$, meaning that H is not a member of the set.

Similarly,

$$14.2 \in \{h \mid 0 < h < 15\}$$

But

$$15 \notin \{h \mid 0 < h < 15\}$$

because 15 is not less than (<) 15.

Consider another example: with the proposition 'It will rain tomorrow in the capital of Borgonia', there are many ways to construct the domain. What is the exact specification of the time period ('tomorrow')? What is the exact specification of the geographical area covered? What constitutes the event 'It is raining'? These questions can be answered if it is decided unambiguously in advance which evidence will be agreed by everyone involved as to the truth of the proposition. If you were to take up a gambling bet with an opponent, these things would have to be unambiguously specified in order to avoid arguments about who won the bet.

Events

Let's consider the domain of two-coin tosses $\{HH, HT, TH, TT\}$. The members of this set are the 'elementary events' according to our discussion above. In which types of events might we be interested? Consider the following examples, where on the left-hand side the event is stated in words and on the right-hand side the corresponding set is shown (Table 1.1).

Table 1.1 Some events resulting from two-coin tosses

1. Both coins have the same outcome	$\{HH, TT\}$
2. There is at least one H	$\{HT, HH\}$
3. The two outcomes are different	$\{HT, TH\}$
4. There are two Hs	$\{HH\}$
5. There are twice the number of Hs than Ts	$\{\}$
6. There is at least one H or one T	$\{HH, HT, TH, TT\}$
7. There is exactly one H OR no Hs	$\{HT, TH, TT\}$
8. There is at least one H AND both are the same	$\{HH\}$
9. There is no H	$\{TT\}$

Notice that event 5 is an impossible event. It corresponds to a set that is 'empty' (it has no members). The empty set is usually denoted by ϕ (the Greek letter 'phi').

Now consider some examples for heights, with a person selected at random from a population with domain $\{h \mid h > 0\}$ (Table 1.2).

Table 1.2 Some events resulting from measuring the height of a person

1. The person is taller than 1.68 m	$\{h \mid h > 1.68\}$
2. The person is shorter than 1.58 m	$\{h \mid h < 1.58\}$
3. The person is between 1.58 and 1.68 m	$\{h \mid 1.58 < h < 1.68\}$
4. The person is exactly 1.68 m tall	$\{1.68\}$
5. The person is taller than 1.68 m and less than 1.58 m	ϕ
6. The person has a height	$\{h \mid h > 0\}$
7. The person is shorter than 1.58 m OR taller than 1.68 m	$\{h \mid h < 1.58 \text{ OR } h > 1.68\}$
8. The person is at least 1.58 m AND at most 1.58 m	$\{h \mid h \leq 1.58 \text{ AND } h \geq 1.58\} = \{1.58\}$
9. The person is NOT taller than 1.68 m	$\{h \mid h \leq 1.68\}$

Do you notice something common about all the events expressed as sets? They are all *subsets* of the domain. A subset means that it is *part of* the set representing the domain. For example, considering event 1, $\{HH, TT\}$ is part of $\{HH, HT, TH, TT\}$, and $\{h \mid h > 1.68\}$ is part of $\{h \mid h > 0\}$. It is the same for all of the other events. Of course, the individual elements of the domain, considered as sets, are subsets of the domain as in example 4. By convention, the empty set (the set with no elements at all) is always considered as a subset (example 5), and the domain itself is considered a subset of itself (example 6). The empty set is an 'impossible event' and the domain itself is the 'certain event' (it must happen).

Generally, if S is any set and V is a subset of S, then we write $V \subset S$. Of course, V itself is a set. The domain is usually denoted by the symbol Ω (the Greek capital letter 'omega'). So any event E is such that $E \subset \Omega$.

In general, if A and B are sets, then the *union* of the two sets consists of all the elements of one and all the elements of the other pooled together (if they have elements in common, these are only included once). This is denoted by $A \cup B$. In terms of events, the event $A \cup B$ corresponds to A OR B. In the coin example 7 (Table 1.1),

'There is exactly one H' OR 'no Hs' $= \{HT, TH\} \cup \{TT\} = \{HT, TH, TT\}$

For the heights example 7 (Table 1.2),

'The person is shorter than 1.58' OR 'taller than 1.68' $= \{h \mid h < 1.58\} \cup \{h \mid h > 1.68\}$

If A and B are sets, then their *intersection* consists of the elements that are in common between them, denoted by $A \cap B$. For events, intersection is equivalent to AND, both A and B occur or are true. Note that the intersection might be empty (impossible) as in example 5. Consider example 8 for the coins (Table 1.1):

'There is at least one H' AND 'both are the same' = $\{HT, HH\} \cap \{HH, TT\} = \{HH\}$

Consider example 8 for the heights (Table 1.2):

'The person is at least 1.58' AND 'at most 1.58 m' = $\{h \mid h \geq 1.58\} \cap \{h \mid h \leq 1.58\} = \{1.58\}$

In this case, 1.58 is the only element in common between the two sets.

If A is a set (a subset of the domain Ω) then $\Omega - A$ is the set that has all the elements of Ω which are not in A. This is usually called the *complement* of A. In terms of events, this is 'not-A' or 'A is not true'. Consider example 9 for the coins (Table 1.1):

'There is not a H' = $\{HH, HT, TH, TT\} - \{HH, HT, TH\} = \{TT\}$

For the heights example 9 (Table 1.2),

'The person is NOT taller than 1.68' = $\{h \mid h > 0\} - \{h \mid h > 1.68\} = \{h \mid h \leq 1.68\}$

Some final terminology in this section: a set can have elements that are themselves sets. For example, $\{\{HH\}, \{HH, TT\}, \{HT, TH, TT\}\}$ is a set with each element itself a set. Similarly, $\{\{h \mid h > 158\}, \{h \mid h < 168\}, \{h \mid 1.58 < h < 1.68\}$ is a set with three elements which are themselves sets. In this case, the set is called a *class*. There is nothing new here, it's just a name. Thus we can talk about the *class of all subsets of a set* Ω, which is a set consisting of all the subsets of Ω. More about this below.

Axioms of Probability

To unify terminology for events and propositions, we will adopt the terminology 'the event is false' to mean that it did not, or will not happen, and similarly 'the event is true' to mean that it did or will happen.

The domain as we discussed above is denoted by Ω, the set of all possible 'elementary events'. Let $\mathfrak{I}$ be a class of subsets of Ω, with the following properties:

A1. $\phi \in \mathfrak{I}$.
A2. $\Omega \in \mathfrak{I}$.
A3. *If* $E_1, E_2, \ldots, E_n \in \mathfrak{I}$, *then the union of all of them*, $E_1 \cup E_2 \cup \ldots \cup E_n \in \mathfrak{I}$, *or* succinctly, $\cup_{i=1}^{n} E_i \in \mathfrak{I}$.
A4. If $E \in \mathfrak{I}$, then $\Omega - E \in \mathfrak{I}$.

For example, in the case of tossing two coins, we would have

$\Omega = \{HH, HT, TH, TT\}$

Examples of subsets of Ω are $E_1 = \{HH\}$, $E_2 = \{HH, HT\}$, $E_3 = \{HH, HT, TH\}$. Note that Ω as we pointed out is considered as a subset of itself and always the empty set $\phi = \{\}$ is a subset. In this particular case,

$$\Im = \{\phi, \{HH\}, \{HT\}, \ldots, \{HH, HT\}, \ldots, \{HT, TH, TT\}, \ldots, \{HH, HT, TH, TT\}\}$$

There will be 16 elements in $\Im$ (the empty set, all singletons, all pairs, all triples, and all four together). Notice, of course, that our example E_1, E_2, E_3 are members of $\Im$.

To illustrate A3, where $\cup$ denotes *union*, considering our example E_1, E_2, E_3,

$$E_1 \cup E_2 \cup E_3 = \{HH, HT, TH\}$$

which is also a member of $\Im$.

To illustrate A4,

$$\Omega - E_1 = \{HH, HT, TH, TT\} - \{HH\} = \{HT, TH, TT\}$$

which is also a member of $\Im$.

In this notation, $\Im$ is the class of events to which probabilities may be validly assigned. The first requirement (A1) is that the empty set be part of this class. The 'empty set' corresponds to an event that is *impossible*. In a sense, all impossible events, whatever they refer to, are the same – they are impossible. So just a single notation is required to stand for the impossible event, equivalent to the empty set.

It is important to note that *impossible* here is meant in a strict sense. Effectively, it is logically impossible, it cannot possibly be true: $1 + 1 = 3$. However, an event such as you keep tossing a coin 'for ever' and it always shows Heads – that is in principle a possible event. It is an event that is not impossible, but nevertheless has 'probability 0' (we will come back to that).

The second requirement (A2) is that the 'certain event' (i.e. 'something happens') is a member of the class. For example, in the example of heights, $\Omega = \{h|h > 0\}$. The event Ω is 'the height is a positive number'. This is certain.

The third requirement (A3) is that the union of events, equivalent to the proposition 'at least one of the E_i occurs', is a member of the class, provided that each of its constituent events are in the class. For example, 'the height is greater than 1 m' and 'the height is less than 1.5 m' have the union 'the height is greater than 1 m or the height is less than 1.5 m'. Remember that the symbol $\cup$ (union) is interpreted as 'OR'. The first two are valid events, and, therefore, so is their union.

The fourth requirement (A4) is that if an event E is part of the class, then its complement 'E is not true' must also be part of the class. For example, if E is the event 'the height is greater than 1 m', then the event 'the height is less than or equal to 1 m' is also an event.

Such a class of subsets is called a *sigma field*. We are stating this for interest, or in case the reader carries on to more advanced courses in probability; we will never use this again in this book. In the axiomatic treatment of probability, the triple $(\Omega, \Im, P)$, consisting of the domain Ω, the events $\Im$, and a probability measure over the events P, is the basic mathematical object of investigation.

Probability theory is mathematically founded on a set of *axioms*. Axioms can be considered as self-evident 'truths' or statements that require no additional proof. The entire theory of probability is founded on axioms proposed by A. N. Kolmogorov in 1933. An interesting history of probability can be found in (Debnath and Basu, 2015). The measure P satisfies a number of *axioms*:

A5. For any event E, $P(E) \geq 0$.
A6. $P(\Omega) = 1$.
A7. For any events $E_1, E_2, \ldots, E_n$, *where only one of these can occur*, then the probability of one of them occurring is the sum of their probabilities. The stipulation that only one of them can occur means that for any two of them, E_i and E_j, $E_i \cap E_j = \phi$ (they both cannot be true, their intersection is empty).

Writing this more formally,

$$P\left(\cup_{i=1}^{n} E_i\right) = \sum_{i=1}^{n} P(E_i) \text{ provided that } E_i \cap E_j = \phi \text{ all } i \neq j.$$

A5 says that probabilities are at least 0. A6 says that the probability of 'the certain event' is 1.

To recap A7, we have a requirement on the events. The probability of a *union* of events, *where one and only one of these events is true*, is equal to the sum of their probabilities. Such events are called *mutually exclusive*.

It follows from this that $P(\phi) = 0$ (an impossible event has probability 0). This is because $\phi \cup \Omega = \Omega$, and $\phi \cap \Omega = \phi$. Therefore, from A6: $P(\phi \cup \Omega) = P(\Omega) = 1$.

However, from A7, $P(\phi \cup \Omega) = P(\phi) + P(\Omega)$. Therefore $P(\phi) + P(\Omega) = 1$ so that $P(\phi) = 0$.

The notation $\bar{E} = \Omega - E$ is used to denote the complement of an event, that the event that E does not happen, or that the proposition E is false.

From A5 through A7, it is easy to show that $P(\bar{E}) = 1 - P(E)$. In other words, the probability of an event being false is one minus the probability of its being true. Can you show this?

Here it is:

Since E and $\bar{E}$ cannot both be true, $E \cap \bar{E} = \phi$. So from A7,

$$P(E \cup \bar{E}) = P(E) + P(\bar{E})$$

But $E \cup \bar{E} = \Omega$ (it is certain that either E or its complement will occur!).

Therefore from A6, $P(E \cup \bar{E}) = 1$, so that

$$P(E) + P(\bar{E}) = 1$$

From this it also follows that for any event E that $P(E) \leq 1$, for, if not, then this would violate the requirement that $P(E) \geq 0$.

Conditional Probabilities and Independence

All probabilities are really conditional. For example, the event 'It will snow some time next week somewhere in Borgonia' is conditional on Borgonia, even the world, still existing. It would be tedious to keep saying 'conditional on the world still existing ...'; for any event, there are infinite unconsciously accepted conditions behind that event. 'Conditional on the world still existing in the next 10 seconds, and conditional on my actually having a coin, and conditional on my hand not getting stuck in my pocket, and ... the probability of this coin landing Heads uppermost when I get round to tossing it after this long statement is ...' whatever it will be.

So, in this book we will not always state the conditions, otherwise the book could become exceedingly lengthy, and without giving any additional knowledge that everyone wouldn't have already assumed.

This is the point: when there are conditions that *should* be stated explicitly, we use the notation $P(E \mid H)$ to denote 'The probability of E conditional on H'. If you knew H to be true, what is your probability for E? If you had a long-range weather forecast, your probability for 'It will snow some time next week somewhere in Borgonia' should be different, compared to knowing nothing whatsoever about the Borgonia weather forecast. Of course, *since probabilities are assigned subjectively*, you can choose whatever you like. However, the theory of probability will give you rational ways to change your probabilities in the light of additional information. By rational is meant that the mathematics of probability derived from the axioms will give you ways to assign probabilities to more complex events given probabilities of more constituent events. For example, if you had assigned the probabilities to 'it will rain all day in Borgonia tomorrow' and 'it will snow all day in Borgonia tomorrow' then the event 'it will rain all day or snow all day tomorrow in Borgonia' can be assigned a probability from A7 (assuming that it cannot rain *and* snow).

All probability statements have a silent and implicit 'conditional on X'. However, when X is obviously understood by all concerned, it is tedious and cumbersome to have to write it.

Conditional probability is involved in the final axiom:

A8. $P(A \cap B) = P(A|B)P(B) = P(B|A)P(A)$

This states that the probability of events A and B being both true may be obtained from the product of the conditional probability of A given B and the probability of B (and similarly, swapping the roles of A and B where this makes sense).

It will sometimes happen that A and B do not depend on one another. For example, the outcome of the coin toss and the snow in Borgonia are unlikely to be related (ignoring chaos theory). In this case, the conditional probability is the same as the probability, and the probabilities simply multiply. The probability of A and B being simultaneously true is then the *product* of their probabilities. In this case, where A and B do not depend on one another at all, then from A8,

$$P(A \cap B) = P(A|B)P(B) = P(A)P(B)$$

since $P(A|B) = P(A)$ and $P(B|A) = P(B)$.

This expands into the following *definition* (it is not an axiom but introduces a terminology):

A9. The events $E_1, E_2, \ldots, E_n$ are *independent* if and only if the probability of them all happening (all being true) is the product of their probabilities. Formally, we can write

$$P(E_1 \cap E_2 \cap \ldots \cap E_n) = P(E_1)P(E_2) \ldots P(E_n)$$

or using a shorthand,

$$P\left(\bigcap_{i=1}^{n} E_i\right) = \prod_{i=1}^{n} P(E_i)$$

and that this holds for every possible subset of $E_1, E_2, \ldots, E_n$. Here the operator

$$\bigcap_{i=1}^{n} E_i$$

is just a convenient way of writing $E_1 \cap E_2 \cap \ldots \cap E_n$, and

$$\prod_{i=1}^{n} P(E_i)$$

is a shorthand for $P(E_1)P(E_2) \ldots P(E_n)$.

For example, for events A, B, and C to be independent, we must have

$$P(A \cap B \cap C) = P(A)P(B)P(C)$$

$$P(A \cap B) = P(A)P(B)$$

$$P(A \cap C) = P(A)P(C)$$

$$P(B \cap C) = P(B)P(C)$$

and, of course, the singletons $P(A) = P(A)$ are obvious, and so is the empty set $P(\phi) = P(\phi)$.

By the way, this rule allows us to reconsider the apparent paradox that there can be theoretically possible events that have zero probability. In A9, let every E_i be the event 'toss a coin and get Heads'. Successive tosses are *independent*. Each $P(E_i) = 1/2$. So for any n, the probability that all tosses result in H is $(1/2)^n$. It is clear that for larger and larger n, this probability is smaller and smaller. In the theoretical limit of $n \to \infty$, the probability becomes 0. Hence, here is an example of a theoretical event that is not logically impossible, but which nevertheless has probability 0. You sometimes might come across phrases such as 'Event E will occur with probability 1', where E is not the *certain* event. Just as an event can be theoretically possible but have probability 0, so it may not be a logically certain event, but have probability 1. Can you think of an example?

On another point if you think about it, there was nothing special about every toss resulting in a Head. Consider this: a coin is tossed 100 times. What is more likely, all Heads or 50 Heads and 50 Tails in the sequence HTHTHT ... HT?

Your intuition may tell you that the interspersed sequence of 50 Heads and 50 Tails is more likely than all Heads. But apply the same reasoning as above. The probability of the sequence HTHT … HT is also $(1/2)^n$ (where $n = 100$). Our statistical intuition often fails. There are many fascinating examples of this in Kahneman (2011). In this case, we end up with the strange situation that if we toss a coin a large number of times, then every possible outcome has near 0 probability of occurring (as near as 0 as you want, by choosing n large enough).

Assigning Probabilities

Now that we know the axioms, we still need a starting point. How are probabilities assigned at all? Probability theory allows us to manipulate probabilities, and assign probabilities to more complex events given that we know probabilities of simpler events, but we have to have a 'bootstrapping' operation to get started – we have to have something that lifts us out of the abstraction into concrete assignment of probabilities. There is no 'correct' answer to this. However, there are three common approaches.

Equal Probabilities

For any domain, if you 'know nothing' assign equal probabilities. The two-coin toss is an obvious example. The domain is $\{HH, HT, TH, TT\}$, and so each of these events would be assigned a probability of 1/4 in the absence of any prior information.

The event (A) 'there will be at least one Head' corresponds to the subset $\{HH, HT, TH\}$ and so would have probability 3/4. The event (B) 'both coins will be the same' corresponds to the subset $\{HH, TT\}$ and so has probability 1/2.

What is $P(A \cup B)$? We cannot invoke A7, since A and B are *not* mutually exclusive. In such a case in general, we can argue as follows:

$$A \cup B = A \cup (B - A)$$

(If you join A and B together, then this is the same as joining A with all the elements in B that are not already in A.)

By A7, since the terms on the right-hand side are mutually exclusive (it cannot be the case that both of A and $B - A$ are true),

$$P(A \cup B) = P(A) + P(B - A)$$

In addition,

$$(B - A) \cup (A \cap B) = B$$

(If you join together all the elements of B that are not in A, with all the elements in both A and B, then you are left with only the elements of B).

The left-hand side of this equation consists of two mutually exclusive events, and therefore, $P(B - A) = P(B) - P(A \cap B)$.

Putting all this together, we have the new finding:

A10. $P(A \cup B) = P(A) + P(B) - P(A \cap B)$

Note that this is not an axiom, since it is derived from them. It is an example of how we can rationally assign probabilities to more complex events if we know the probabilities of their constituent simpler events.

Now the event that (*A*) 'there will be at least one Head' and event (*B*) 'both coins will be the same' correspond to {*HH*}, which has probability of 1/4. Therefore, the probability of *A* or *B* occurring is (3/4) + (1/2) − (1/4) which is 1. Now this would be an interesting proposition on which to bet with a probabilistically naive opponent.

Here is an interesting example of the use of equal probabilities with a surprising outcome. How many people (n) are needed so that the probability of at least two of them having the same birthday exceeds 1/2? Let's simplify and consider 365 days in a year, ignoring leap years. Starting from any arbitrary person (1), the probability that another person (2) will have a different birthday is 364/365 (assuming equal probabilities for all the days of the year). The probability that a third person will have a different birthday to the first two is 363/365. Continue on to the nth person, and the probability that no two of them have the same birthday is

$$\left(\frac{364}{365}\right)\left(\frac{363}{365}\right)\cdots\left(\frac{365-n+1}{365}\right)$$

The probabilities must be multiplied since here we want the probability that no two of them have the same birthday, and assuming independence (e.g. no twins). We can calculate this with a simple R function:

Box 1.1

```
birthday <- function(n) {
#the probability of no 2 people amongst n having the same birthday
#assuming equal probabilities of the 365 days.
   p <- 1
   for(i in 2:n) {
      p <- p*(365-i+1)/365
   }
return(p)
}
```

The function 'birthday' computes the probability of there being no two people with the same birthday amongst n. We find that birthday(10) = 0.88, so the probability of there being at least two with the same birthday is 0.12. However, birthday(22) = 0.52, and birthday(23) = 0.49. So with only 23 people, the probability is more than 0.5 that there are at least two with the same birthday; birthday(30) = 0.29, so 0.71 is the probability that there are at least two with the same birthday. For more information, see ScienceBuddies (2012).

Betting Odds

'Odds' are simply ratios of probabilities. They are important for judging which of two events is more likely, without having to know the actual probabilities of the two events. For example, suppose

$$\frac{P(A)}{P(B)} = 3$$

Then this gives us the information that A is three times more likely than B. Of course, we don't know what these probabilities are. For example, both of these would be valid:

$$P(A) = 0.003,\ P(B) = 0.001$$

and

$$P(A) = 0.9,\ P(B) = 0.3$$

In the first case, neither A nor B are likely to be true, but in the second case, A is very likely to be true, and more than B. But in either case, if we *had* to choose between A and B, the choice of A would be wiser.

In everyday life, betting odds are a particular example of odds and have very practical implications. Betting odds are simply ratios of probabilities, and therefore given betting odds, it is possible to work backwards to get the probabilities. Suppose that an 'opponent' offers odds of 10 to 1 if the horse 'Borgonian' wins the next race. This means that for every \$ or € (for example) that you bet, you would get back 10 if Borgonian won the race, otherwise you would lose the money that you bet. Putting this in probabilistic terms, the bookmaker, the one who sets the betting odds, either keeps your bet if you lose or pays out your winnings, and has implicitly assigned a probability of 1/11 to Borgonian winning, and 10/11 to that horse losing. In accepting the bet, you have agreed with these assignments. In general, if someone has assigned $p(E) = p$, and $q = 1 - p$, then their odds in favour of E are q/p to 1. Rationally, you would only agree to a bet with them, on these odds, if you had assigned the same probability.

Betting is a useful way to assign probabilities, since it involves real risk, and therefore forces the assignee, assumed to be behaving 'rationally', to assess the costs and benefits involved in taking on the bet at a given set of odds. In theory, the rational and fair choice is to choose odds where the expected gain exactly balances the expected loss, where gain and loss are measured in terms of the 'utility functions' of the people concerned (Fishburn, 1968). Utility functions theoretically take into account all that the person knows and values and owns, with the abstract quantity that economists call 'utility' applied to the outcomes. Of course, this does not take into account power relationships, where the person choosing the odds has the power to impose them on you, but that is another story. There is a whole subject called 'decision theory' that takes into account utilities of outcomes and their probabilities and argues about how to make rational decisions in the light of this information (Peterson, 2017).

Frequency

Let's suppose, though, that utility simply boils down to money. Suppose you're betting again and again with a worthy and fair opponent. Week after week, you play a chess game with that person, and you bet on who wins. You have observed that over a long period, the ratio of the number of times you have won to the total number of games has stabilized at the proportion p. For example, after the first 20 games, you noticed that you had won eight times. Over the first 50 games, you had won 22 times. After the first 100 games, you had won 38 times. Over the past 500 games, you had won 195 times. The long-term proportion seems to be stabilized at around 0.4. In these circumstances, what are the 'odds' you would agree to for the event that you will win the next game?

Suppose you choose odds of 3 to 1 that you will win. Then using these odds over past data, your net gain per unit bet would be $0.4 \times 3 - 0.6 = 0.6$. In other words, 40% of the time you would have gained 3, and 60% of the time you would have lost 1. This is great for you, but your opponent would think that these were not fair odds. On the other hand, suppose that the odds were 'evens' (1 to 1). Then your net gain would be $0.4 - 0.6 = -0.2$. Your friend would be laughing. It is clear that the only 'fair' odds, odds that do not give any special advantage to either player, are 3:2. Then your net gain is $0.4 \times 3 - 0.6 \times 2 = 0$, and this therefore gives each player the same long-run outcome. What the actual outcome is, of course, down to 'luck' – sometimes one will be gaining and sometimes the other.

In general, if such frequency information is available, then fair odds are q/p to 1, where $q = 1 - p$, and therefore the probability is p. This is so because the expected gain is

$$(q/p) \times p - q = 0$$

(proportion p of the time, you will win q/p, and proportion q of the time you will lose 1).

It should be clear from this discussion that although you are completely free to assign any probability you like to an event, if it is to have any impact on the world, then it is likely to be a probability that is agreed with others. Probability assignment, or assignment of odds, is a transaction between people, a social event. Fair players would agree on odds in relationship to long-term frequencies, where these are relevant. In real life, power and economic relationships distort this.

By the way, this is one way how casino owners make their money. They give odds of 'evens', for example, on the bet of 'red' or 'black' in roulette. But of the 37 slots on the roulette table, one is zero, which is neither red nor black. They pay odds of 35 to 1 on any individual number. Fair odds would be 36 to 1.

Now the argument about long-term proportions rests on many assumptions (unlikely to be realized in practice). Going back to the chess game, one is that each game is played under 'identical conditions'. Of course, sometimes you will feel great, and other times, you'll have a headache, and the same will be true for your opponent. The cold weather might affect your brain, and the warm weather your opponent's brain. Second, it assumes 'independence' between all the games – that is, that the outcome of one game cannot influence the outcome of any other game. This is unlikely to be the case since strategies will develop over time, and there will be learning. Loss of a particular game might make you especially determined to win the next one and therefore spend time reading up on chess strategies. Hence, assigning probabilities through long-run frequency, even when such data may be available, is not always advisable, since non-independence between outcomes may greatly distort the frequencies.

It used to be the case that the only widely accepted definition of probability was this 'long-run frequency'. In this folklore, a hypothetical experiment is repeated n times under identical conditions and with independence between the trials. If event E happened on r of the n occasions then an approximation to $P(E)$ would be r/n. As n tends to infinity, this ratio approaches the 'true probability' of E. To use our previous terminology, the Law of Large Numbers states that under these conditions $\frac{r}{n} \rightarrow p$ *with probability 1*, where p is the 'true probability'. ('With probability 1' means that it is theoretically but not practically *possible* that the limit might not be p.)

This frequency definition rules out the assignment of probabilities to most events or propositions. We could not talk, for example, of the probability of the horse Borgonian winning the next race, because there has never been a sequence of races under identical conditions (same opponent horses, same day, same weather, same place). Polling forecasters talk about the probability of a particular candidate X winning an election. In the frequentist approach, such probability assignments are meaningless, since the argument would have to be like this: in multiple parallel universes, candidate X is standing for election under identical conditions. The proportion of universes in which X wins is the probability of X winning. 'Amongst n possible universes where n is large, X won in r of them, and therefore the probability that X will win in this universe is r/n.' If the physics theory of the multiverse is true, maybe this makes sense. But unfortunately, those other universes are currently unobservable.

Summary

In this chapter, we have introduced the idea of probability based on a triple of domain, the class of events, and a measure obeying a number of axioms. We have interpreted probability as a number between 0 and 1 (inclusive) that is subjectively assigned representing the assignee's 'degree of belief' in the event being 'true'. If an event is certain, its probability is 1, and 0 if it is impossible. We have argued that probability theory allows the rational assignment of probabilities, following the mathematical rules of probability derived from the axioms, to more complex events given assignments to simpler events. We considered three ways to assign probabilities to simpler events: by the argument of equal probability in the absence of prior information, by betting odds, and by long-run frequency information if this exists.

In Chapter 2, we will use the rules of probability to derive probabilities over number spaces: the concept of 'random variables'. Random variables are at the heart of statistics, since the data that we collect are observations on such random variables. Going back to the beginning of this chapter, when we select a 16- or 20-year-old in order to measure their height, the measurement of height yields an observation on the random variable 'height'.

Probability theory is a fascinating subject, with many counter-intuitive surprises that can be derived from the theory. An excellent, if maybe difficult to get-hold-of, book is *An Introduction to Probability Theory and Its Applications* by Feller (1957).

Online Resources

The birthday example is programmed in R and can be accessed at

www.kaggle.com/melslater/slater-bayesian-statistics-1

Two

PROBABILITY DISTRIBUTIONS

Introduction

It is particularly interesting when an event is expressed in quantitative terms. For example, we can consider the event 'Political Party X will win the next general election in Borgonia'. We can also consider the event 'The government formed after the next general election in Borgonia will have a majority of n in the Parliament' where n is an integer in the appropriate range. If a probability were assigned to each possible value of n, then this would be the *probability distribution* for n. Before continuing, it is worth noting that this chapter introduces some mathematical notation. There are many books that could help you brush up on your mathematics knowledge: for example, take a look at Hagle (1995, 1996) and Kropko (2015).

A *random variable* is just such a quantified event. It can take values in some given range, which may be infinite or finite, continuous or discrete. A finite range is one that is determined by two specific numbers, for example between 0 and 1, or between –124 and 1120. An infinite range is one where at least one of the bounds is infinite. For example, if we say that the range consists of all numbers less than 10, then the lower bound is minus infinity, $-\infty$. If we say that the range consists of all non-negative numbers then the range is all numbers that are 0 or greater, so that the upper bound is ∞. If the range is any number at all, then it is from $-\infty$ to ∞. A discrete variable is one where the possible values can only range over steps: for example, 1, 2, 3, ... or 5, 7, 9, ...; in general, it has the form $a + db$, where a and b are fixed numbers, and, for example, $d = \ldots, -2, -1, 0, 1, 2, \ldots$. A discrete variable may also have an infinite range. A discrete variable may be over a sequence of numbers with no particular structure, though this is unusual. A continuous variable is one that can vary continuously, for example *all possible numbers* in the range 0 to 1. A continuous variable can have a finite range (e.g. 0 to 1) or an infinite range (e.g. any positive number, or any number at all from $-\infty$ to ∞).

A random variable has a probability that is associated with every subset of its range. We will see many concrete examples later.

We distinguish between the 'name' of a random variable (e.g. 'the number N of seats that the governing party in Borgonia will have in the Parliament' – N is the name) and the possible actual value (x) that the variable can take (in this case, $0, 1, \ldots, n$, where n is the actual size of the Parliament). We use notation $N \leq x$ to stand for the event that an observed value of the random variable will be less than or equal to some given number x. For example, as is well known, the total number of Members of the Parliament in Borgonia is $n = 300$. We may be interested in the event $N \leq 150$, which is the event that the governing party does not have a majority. This event will have some associated probability. N is the random variable.

In general, suppose that X is some random variable, and that its range of possible values is between a and b ($a < b$), including the possibility that $a = -\infty$ or $b = \infty$, or both. The event $X \leq x$ is of special significance – its consideration leads to the construction of probability distributions. The probability $P(X \leq x)$ is the starting point for this, and it is called *the distribution function* for the random variable. It is generally written as follows:

$$F_X(x) \equiv P(X \leq x) \tag{2.1}$$

Where the context is obvious, that is which random variable is being referred to, it is written simply as $F(x)$. Equation (2.1) gives the probability that the random variable in question (X) takes a value that is less than or equal to some specified value x.

The properties of the distribution function should be evident, following from the axioms in Chapter 1 (prove these results for yourself):

$$F(a) = 0 \tag{2.2}$$

$$F(b) = 1 \tag{2.3}$$

$$F(x + h) \geq F(x) \text{ for all } h \geq 0 \tag{2.4}$$

Equation (2.2) says that it is not possible to find a value of X less than its minimum. Equation (2.3) expresses the fact that every possible value of X is less than or equal to its upper bound. In this context, it is worth remembering that the range of a random variable may be infinite, so that $a = -\infty$ and $b = \infty$ are valid possibilities. Equation (2.4) expresses the requirement for F to be a non-decreasing function. The probability of observing a value less than or equal to x cannot decrease as x increases.

Let's consider a case where the range of values for X is discrete, that is a set of individual values such as whole numbers rather than a continuous range. Without loss of generality, we will assume that this range consists of the set of successive integers $a \ldots b$ where $a < b$ are two integer numbers. Now consider the event $(X = x)$, 'X takes the specific value x', where $x \in \{a, a+1, a+2, \ldots, b\}$ (x is one number in the set $a, \ldots, b$). The event that X is at most x is equivalent to saying that X is at most $x - 1$ *or* X is equal to x. Symbolically,

$$(X \leq x - 1) \cup (X = x) = (X \leq x) \tag{2.5}$$

The union on the left-hand side of Equation (2.5) consists of two *mutually exclusive* events (X cannot be at most $x - 1$ and also x). Therefore, applying rule A7 of Chapter 1:

$$F(x - 1) + P(X = x) = F(x)$$

and so,

$$P(X = x) = F(x) - F(x - 1) \tag{2.6}$$

$P(X = x)$ is called the *probability density function* (or *pdf*) for X, and for a discrete variable may be found from Equation (2.6). (For discrete random variables it is sometimes referred to as the 'probability mass function', pmf). It will usually be denoted by $f_X(x)$ or just $f(x)$, provided that the random variable involved (X) is obvious from the context.

As a simple example, the elevator in the tallest building in Borgonia can carry a maximum of nine people. Suppose on one of your visits there, you are waiting on the ground floor for the elevator to arrive and are speculating about how many people might exit from it. This is a random variable (N) which can take the values 0, 1, 2, ..., 9. Suppose you know nothing whatsoever about the patterns of use of the elevator, so you would assign equal probabilities to each possible outcome. Hence, here the pdf is

$$P(N = n) = \frac{1}{10}, \; n = 0, 1, \ldots, 9$$

This is an example of the *discrete uniform distribution* in the range 0, 1, 2, ..., 9. We will refer to this in subsequent examples. It is called 'uniform' because all the probabilities are the same.

Now consider another type of variable X that varies *continuously* between a and b. The quantity $F(x) - F(x-1)$ now represents the probability $P(x - 1 \leq X \leq x)$; that is, that X takes a value that is between $x - 1$ and x, and there are uncountably infinite such values in this range (or in any range, however small).

A more appropriate way to consider random variables that have a continuous range is to appeal to the calculus. Let $dx > 0$ be the famous 'infinitesimally small' quantity and such that

$$\frac{F(x+dx)-F(x)}{dx} \to f(x)$$

as $dx \to 0$.

Then, writing $dF(x) = F(x + dx) - F(x)$:

$$dF(x) = f(x)dx \tag{2.7}$$

and $f(x)$ is called the *probability density function* for a continuous random variable. Equation (2.7) gives an interpretation for this: $f(x)dx$ is the probability of X taking a value in a 'small neighbourhood' about x.

As an example, suppose that you carry out research at the tallest building in Borgonia because you're interested in the proportion of overseas tourists who visit the building over the course of a month, and you are setting up a study to assess this. Before you have any data whatsoever, you know nothing about this proportion – any value between 0 and 1 would be equally likely. This can be represented by the *continuous uniform distribution* on the interval [0, 1] (i.e. between 0 and 1 inclusive). Here the pdf is

$$f(x) = \begin{cases} 1, & x \in [0, 1] \\ 0, & \text{otherwise} \end{cases} \tag{2.8}$$

Figure 2.1 illustrates this distribution.

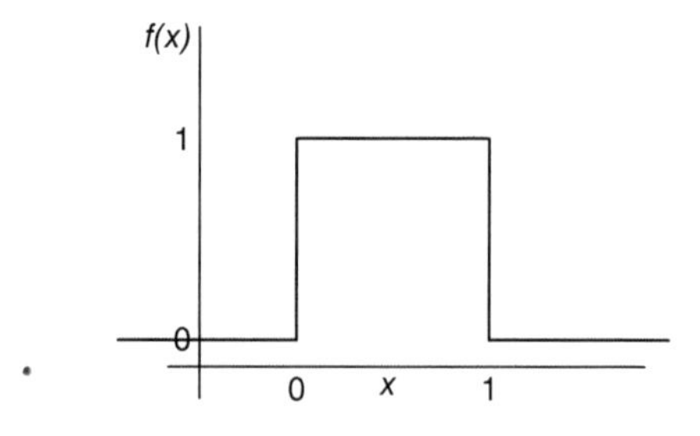

Figure 2.1 The continuous uniform distribution on the interval [0, 1]

This distribution simply expresses the fact that the random variable can take any value in the range 0 to 1 inclusive but cannot take any value outside of this range (because it is a proportion).

The following properties of any pdf follow from the definition, where D is the *domain* of X (the set of values it might possibly take):

- The discrete case, where we consider the sum of $f(x)$ over every possible value x that is in the domain D, that is the sum over every $x \in D$:

$$\sum_{x \in D} f(x) = 1 \tag{2.9}$$

- The continuous case, where we consider the integral over the domain (the area under the curve of $f(x)$ for x in the range given by D):

$$\int_{x \in D} f(x)dx = 1 \tag{2.10}$$

$$f(x) \geq 0 \text{ all } x \in D \tag{2.11}$$

Each of these statements should be checked against the axioms of probability in Chapter 1. By the way, in case you are rusty on this, an integral is the equivalent of a sum, but where the variable concerned is continuous. It is like adding very tiny strips of the function values together to form the area. This is an approximation, but the true area is arrived at as the width of the strips become infinitesimally small.

The pdfs can be used to derive probabilities of interest.

Formally, for the discrete case,

$$P(X \in A) = \sum_{x \in A} f(x) \tag{2.12}$$

which is the probability that the random variable X takes a value in some set A. It is simply the sum of the probabilities associated with each value in the set A.

Returning to the example of the number of people N exiting from the elevator, suppose we wished to know the probability of the elevator being almost empty, say $N \leq 2$. Then, from Equation (2.12),

$$P(N \leq 2) = P(N = 0) + P(N = 1) + P(N = 2) = \frac{1}{10} + \frac{1}{10} + \frac{1}{10} = 0.3$$

For the continuous case (e.g. the proportion of visitors to the tallest building),

$$P(X \in A) = \int_{x \in A} f(x)dx$$

which is the probability that the random variable X takes a value in some set A. The set A must, of course, be a subset of the whole domain D. For example, the probability that the proportion of overseas visitors to the tallest building is less than 1/3 has $A = [0, 1/3]$, the range from 0 to 1/3, which leads to the probability

$$\int_0^{1/3} dx = \frac{1}{3}$$

Another way of expressing this is that *probabilities in the continuous case are equivalent to areas under the curve of the pdf* (recalling that integration gives us the area under a curve). Obviously, the area of the rectangle with base between 0 and 1/3 and height 1 is 1/3, which is the probability. Of course, these examples are very simple, but they illustrate the idea.

Summary Measures of Probability Distributions

In this section, we will consider ways to summarize whole distributions with just a few summary numbers. We will treat the continuous case – the discrete case is the same by replacing integrals with summations (and removing the dx terms). The purpose of the summary measures is to get a 'quick overview' of the distribution, without knowing everything about it. For example, if you are told that the 'average' age of the population of Borgonia is 24 years, this already says a lot about the population of that country, without knowing the entire distribution of ages. If, in addition, you were told that the variation around 24 is 'low', this would tell you even more. If you were asked to guess the age of a person chosen at random from amongst the population, it would be sensible for you to choose 24. On the other hand, had you been told that the variation around 24 was very high, the choice of 24 would not have a high probability associated with it, although it would still be the only thing you could base your decision on if this were all the information that you had.

These ideas can be made precise. The *expected value* of a random variable is just the integral (sum) over all possible values, each weighted by its probability. It is also called the *mean of the distribution* of X. If X is the random variable, with domain D, then the expected value is written $E(X)$ and is defined by

$$E(X) = \int_{x \in D} x f(x) dx \tag{2.13}$$

Let's consider this with respect to the example of the number of people exiting the lift (recalling that this would be a sum rather than an integral). In this case, from (the discrete version) of Equation (2.13),

$$E(N) = \sum_{n=0}^{9} n \times \left(\frac{1}{10}\right) = 45 \times \left(\frac{1}{10}\right) = 4.5$$

This means that on the average if we recorded the number of people exiting the lift many times and then took the average of those numbers, it would approximate 4.5. However, this relies on how well our model that the probabilities are equal (1/10) for each possible outcome corresponds to reality. If our probability model is not a good one to describe what happens in reality, then this estimation of the mean could be way off.

In the continuous case, the proportion of overseas visitors to the building, the expected value is

$$E(X) = \int_0^1 x \, dx = 0.5$$

In other words, in the absence of knowing anything about the overseas number of visitors, expressed through the uniform distribution, the mean value of the proportion is 0.5.

In general, if g is a function over D, then the expected value of $g(X)$ can be calculated as follows:

$$E(g(X)) = \int_{x \in D} g(x) f(x)\, dx \qquad (2.14)$$

For example, with respect to the elevator, $g(n)$ might be the weight of person n. In that case, we would be finding the expected value of the weight of the people exiting the lift. In the case of the proportion of overseas visitors, $g(x)$ could be, for example, their expenditure during their visit to the building. Then $E(g(X))$ would be the expected value of the expenditure.

It is simple to verify that for any constants a and b

$$E(a + bg(X)) = a + bE(g(X)) \qquad (2.15)$$

As was mentioned above, knowing only the expected value is not enough to characterize the distribution. We also need to know how much variation there is: that is, how much on average the observations tend to differ from one another. The *variance* of a distribution is a measure of how much dispersion there is away from the mean (expected value). It is the integral (or sum in the discrete case) of all squared differences between the values and the mean, weighted by the probabilities:

$$Var(X) = E((X - \mu)^2) \qquad (2.16)$$

where $\mu = E(X)$.

Making use of Equation (2.15) then, it is possible to see, by squaring out the brackets in Equation (2.16) and taking the expected value E inside the brackets, that

$$Var(X) = E(X^2) - \mu^2 \qquad (2.17)$$

The variance is measured in squared units. The *standard deviation* is the square root of the variance and, therefore, is in the same units as the original variable. Normally, the standard deviation is represented by the symbol σ, and hence, $Var(X) = \sigma^2$.

Other summary measures that are important for this book are the mode and the median. The *mode* is simply the value of a random variable with the highest probability. Expressing this in terms of the pdf, it is the value corresponding to the maximum of the pdf. (Of course, there may not be a unique maximum – e.g. consider the uniform distribution.) The median is such that if X is a random variable and α is the median, then $P(X \leq \alpha) = 0.5$. So half the distribution lies below and the other half above the median.

There are many other summary measures of distributions, such as skewness and kurtosis, but these do not play a role in this book.

Some Common Distributions

As we will see later, the principles of Bayesian statistics are very easy to grasp based on a basic understanding of probability. However, in order to *apply* Bayesian statistics in a variety of situations, it is important to be familiar with probability distributions that can model both the distribution of the random variable in question (e.g. the proportion of visitors from overseas) and our own uncertainty about the parameters of that distribution such as its mean and standard deviation. For example, if we knew *for sure* that the distribution of visitors followed the uniform distribution on the interval [0, 1], then really there is nothing else to say about this. If we knew this for sure, it would be pointless to collect data, because if it were the case *for sure*, then these data would simply be observations on the random variable X where X has the uniform distribution on [0, 1]. For example, if we collected data on the proportion of overseas visitors for 100 months, and plotted the results in a histogram, we may get something like that shown in Figure 2.2 (compare with Figure 2.1).

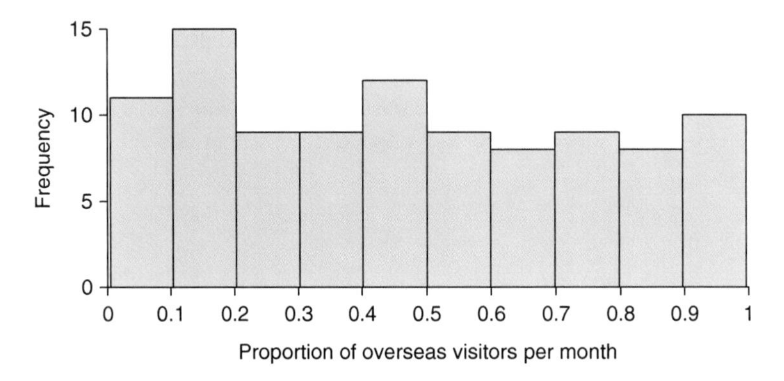

Figure 2.2 Hypothetical histogram of the proportion of overseas visitors per month. Based on 100 observations on a uniform distribution

However, in reality, we would not know the precise probability distribution of the proportion of overseas visitors. Suppose we obtained the proportion of visitors month by month for 100 months and obtained the histogram shown in Figure 2.3. It would seem that the uniform distribution model for these data would be highly inappropriate. In fact, Figure 2.3 is generated from a different distribution called 'Beta', which we will learn about soon. The mean of the observations in this case is 0.24, which is about half of what would be expected in the case of the uniform distribution we have been using as a model. So this is a different distribution, and we would not know the values of the *parameters* (e.g. mean and standard deviation) describing this distribution. A fundamental problem that statistical inference allows us to solve is to estimate these parameters based on the observed data combined with any prior knowledge that we might have had. This issue is covered extensively in the later chapters.

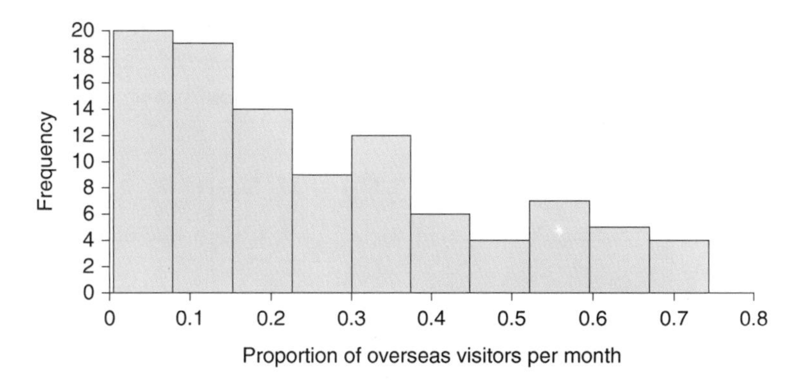

Figure 2.3 Hypothetical histogram of the proportion of overseas visitors per month based on 100 observations from a Beta distribution with parameters (1, 3)

Probability distributions are therefore at the heart of statistics: first, to model the variation in the variables of interest (e.g. the numbers of people exiting the elevator, the proportion of visitors who are overseas), and second to specify our own uncertainty about this model utilizing any prior information we might have. In the first of these, we are using probability to model something about physical reality. In the second, we are representing our own uncertainty. It is critical to understand that both are abstractions. A model is never 'true', it is just an abstract representation, ideally the best representation that we have. Consider again the proportion of overseas visitors. It might actually be the case that month by month the proportion of overseas visitors doesn't vary much. In this case, the choice of the uniform distribution would be fine. However, it is possible that the proportion of overseas visitors peaks in the summer months, to take advantage of Borgonia's fabulous beaches, and at Christmas time, to take advantage of the fantastic shopping offers. In that case, the choice of the uniform distribution to model the proportion of overseas visitors would be inappropriate. So which other possibilities are there?

Next, we will describe the major probability distributions that are typically used in statistical inference. These are usually specified by their pdf. These are given by their functional form (involving algebraic conventions). We begin with the fairly mathematical description of several distinct *univariate* distributions (distributions of a single random variable) followed by consideration of further properties of distributions, which involve more than one variable (*multivariate*) and necessary tools like vector notation in order to conclude the chapter with the description of further important distributions. If you have already completed a basic course in statistical inference, you will probably have heard of and used some the distributions that are covered, namely the normal (or Gaussian) distribution, Student's t distribution, the F distribution, and the multivariate normal distribution. Other listed distributions will appear to be more exotic but almost all will make an appearance in later chapters. In fact, the next sections could be browsed, and then referred to when they make their appearances later.

We start with the simplest distributions and gradually include more and more sophisticated ones. We present the distributions by their functional form, since at the end of the day, there is no better way to understand them. Where possible, we derive the form of the distribution from basic principles, but it is not necessary to follow the mathematics involved – it is more important to understand the circumstances in which these distributions provide useful models.

The Discrete Uniform Distribution

Suppose X is a random variable with domain $\{1, 2, ..., n\}$. If each one of these n values are equally likely, then

$$f(x)=\frac{1}{n},\ x=1,2,\ldots,n \tag{2.18}$$

Note that the 1, 2, ..., n may simply be labels representing n distinct possibilities. Here we assume that the numbers have an intrinsic meaning (like the number of people exiting the elevator). From Equation (2.13), we can find

$$E(X)=\sum_{r=1}^{n}\frac{1}{n}r=\frac{n+1}{2} \tag{2.19}$$

Additionally, using Equation (2.17),

$$Var(X)=\frac{n^2-1}{12} \tag{2.20}$$

(Why is the variance 0 when $n = 1$?)

The Borgonian language and alphabet are unusual. For example, in English the letter 'e' is likely to occur in a piece of text with greater frequency than any other letter. But it seems that in Borgonian, all 21 letters of the alphabet occur with approximate equal frequency. There are no vowels or consonants, at least officially, so that all letters have equal status with respect to their usage. This is true for both the spoken and written language. Like the Hebrew language, each of the 21 letters has a corresponding numerical value (1 to 21 – there is no 0 in Borgonia), which has been the basis of much mystical speculation. The Borgonian Annual Statistical Survey (BASS) always includes a section on language trends. Last year, the observed frequency of each letter in a substantial amount of text sampled from national newspapers revealed the sample distribution shown in Table 2.1 and Figure 2.4. The text consisted of 212,121 letters.

The distribution shown in Figure 2.4 is as we would expect from the discrete uniform distribution. Recall from Equation (2.19) that the theoretical mean of the distribution is (for $n = 21$) 11. The sample mean of the data used to generate the frequency table is $\bar{x}=11.003$. The theoretical variance from Equation (2.20) is $\sigma^2=\frac{21^2-1}{12}=36.67,\ \sigma=6.055$. The variance calculated from the sample is $s^2 = 36.64$, and so the standard deviation is $s = 6.05$. The Borgonians assign great significance to the fact that taking all the letters between $\mu \pm \sigma$ spells the word 'Harmony'.

However, the BASS survey only sampled the national newspapers. Borgonians pride themselves on their poetry. Does poetic writing have a different structure? Unfortunately, Borgonians are also reluctant to reveal their own poetic works, so that it is exceedingly difficult to obtain samples. However, one sample, consisting of only 174 letters, was posted

Table 2.1 Frequency distribution of the 21 letters of the Borgonian alphabet

Value	Count	Percentage
1	10,031	4.73
2	10,077	4.75
3	9,885	4.66
4	10,225	4.82
5	10,226	4.82
6	10,297	4.85
7	10,025	4.73
8	10,179	4.80
9	9,998	4.71
10	10,063	4.74
11	10,012	4.72
12	10,188	4.80
13	10,063	4.74
14	10,121	4.77
15	10,047	4.74
16	10,110	4.77
17	10,317	4.86
18	9,948	4.69
19	10,070	4.75
20	10,090	4.76
21	10,149	4.78
Total	212,121	100

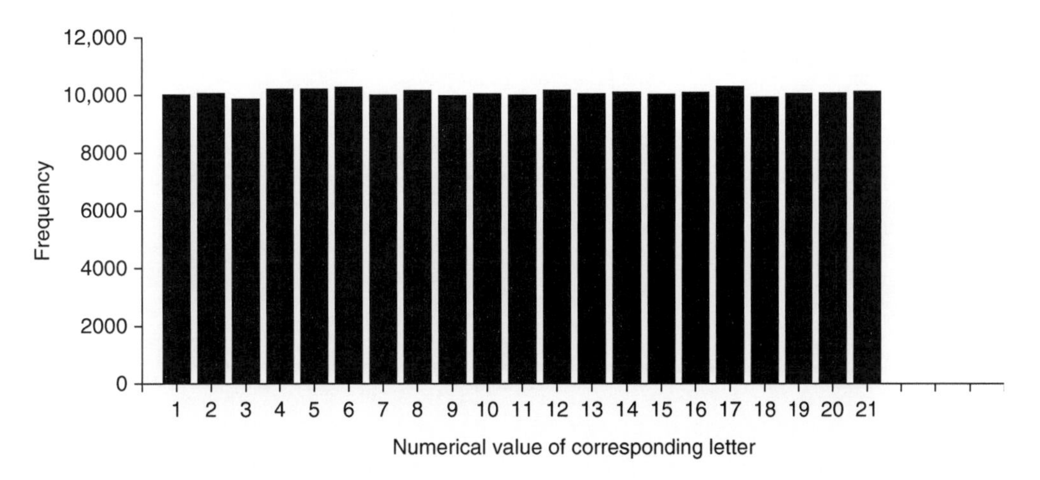

Figure 2.4 Histogram of the occurrences of letters in the Borgonian alphabet

online, anonymously recently, as part of the Borgonian campaign for enhanced literacy. The histogram of the sample is shown in Figure 2.5.

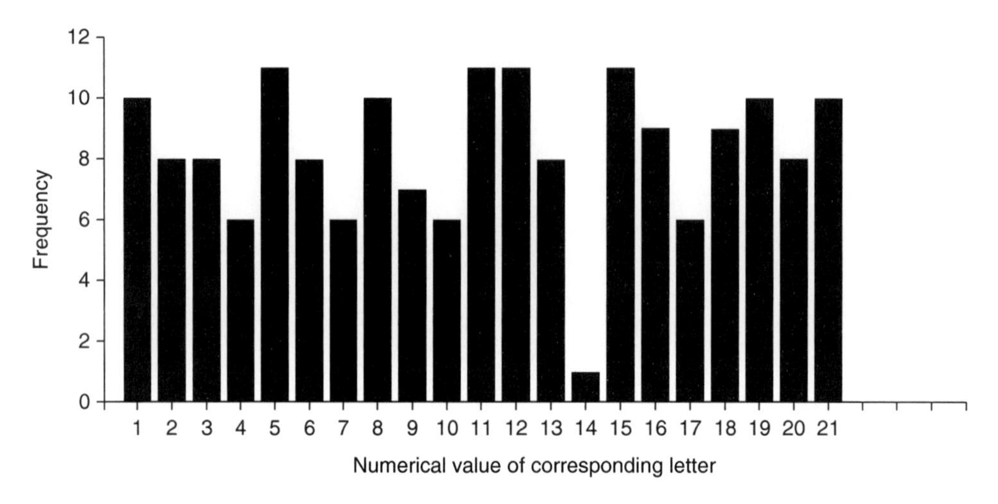

Figure 2.5 Histogram of a sample of 174 letters forming a poem

The sample mean is $\bar{x} = 11.06$ and the standard deviation is $s = 6.2$. The distribution could be from the discrete uniform distribution, and also the sample mean and standard deviation are close to what would be expected from the theoretical distribution. However, we leave such questions of statistical inference to later chapters.

The Continuous Uniform Distribution

In this case, X ranges continuously between the bounds a to b ($a < b$), and 'nothing is known' about X other than this. So the case of the proportion of overseas visitors was a special case of this, where $a = 0$ and $b = 1$. The pdf is

$$f(x) = \begin{cases} \frac{1}{b-a}, & x \in [a, b] \\ 0, & \text{otherwise} \end{cases} \tag{2.21}$$

From this definition, it is straightforward to show that

$$\begin{aligned} E(X) &= \frac{a+b}{2} \\ Var(X) &= \frac{(b-a)^2}{12} \end{aligned} \tag{2.22}$$

For illustrative purposes, we derive the expectation in this particular case. In the later, more complex, distributions, we will not usually derive these values. From Equation (2.16),

$$
\begin{aligned}
E(X) &= \frac{1}{b-a}\int_a^b x\,dx \\
&= \frac{1}{b-a}\left[\frac{x^2}{2}\right]_{x=a}^{x=b} \\
&= \frac{1}{b-a}\left(\frac{b^2-a^2}{2}\right) \\
&= \frac{1}{b-a}\frac{(b-a)(b+a)}{2} \\
&= \frac{a+b}{2}
\end{aligned}
$$

Using Equation (2.17), integrate x^2 following a similar procedure to obtain the result for the variance (Equation 2.22).

We will use the notation $X \sim$ *uniform*(a, b) to denote that the variable X has a uniform distribution on the interval $[a, b]$.

The Binomial Distribution

As the name implies, here we are only interested in two outcomes, whether an *event occurs* or *does not occur* in given circumstances. The probability assigned each time to the *occurrence* of the event is p. The most famous example, of course, is coin tossing. The coin is tossed n times. The event of interest is 'Head'. The random variable of interest is 'how many Heads were there in the n tosses?' where n is fixed in advance. It is assumed that the coin is 'fair' so that each time the probability of Head is 0.5, and also that there is independence between the outcomes. Independence was discussed in Chapter 1.

More generally, an experiment is repeated n times. Each time, an event E may occur or not. The probability for E is p, which is constant. The successive experiments have outcomes that are *independent* in the probabilistic sense. Such trials of an experiment are called *Bernoulli trials* or *Binomial trials*. A 'trial' is what statisticians call repetitions. When you toss a coin repeatedly, the tosses are called 'trials'. They are called Bernoulli trials after the 17th-century mathematician Jacob Bernoulli who introduced these ideas.

Suppose the outcome of trial i is labelled '1' if E occurs and '0' if not. Then the random variable (X) represents the number of times that E occurs, that is the number of 1s. We are interested in the probability of the event $(X = x)$: there are x occurrences of the event E, and $n - x$ occurrences of not-E).

Consider an example sequence of outcomes, where $X = x$:

$$\underbrace{000\ldots000}_{n-x}\underbrace{111\ldots111}_{x} \tag{2.23}$$

This is the particular sequence where the first $n - x$ outcomes are 0, and the next x outcomes are 1. Now if this particular sequence happened, you might be surprised. However, *every such sequence* with x occurrences of E and $n - x$ occurrences of not-E has the same probability, namely $p^x(1 - p)^{n-x}$ (from the rules of independence introduced in Chapter 1).

How many such sequences are there? The answer is the number of rearrangements of the sequence shown in Expression (2.23). This number is given by the combinatorial formula

$$\binom{n}{x} = \frac{n!}{x!(n-x)!}$$

Note that $\binom{n}{x} = \binom{n}{n-x}$.

It is beyond the scope of this book to derive expressions for such combinations, but the reader should try some examples. Suppose there are three objects (A) of one kind and two objects (B) of another. Then one such combination is

AAABB

How many such arrangements are there? According to the formula, it should be

$$\binom{5}{3} = \frac{5!}{2!3!} = \frac{5 \times 4}{2} = 10$$

We have

AAABB, AABBA, ABBAA, BBAAA, BABAA, BAABA, BAAAB, ABAAB, AABAB, ABABA

Can you find any more?

Putting all this together, the binomial distribution for n independent trials where the outcome E has probability p and the random variable of interest is the number of occurrences of E is given by

$$f(x) = \binom{n}{x} p^x (1-p)^{n-x}, \quad x = 0, 1, \ldots, n \tag{2.24}$$

which is the famous *binomial distribution.*

We will write $X \sim$ *binomial(n, p)* to denote that the variable X has the binomial distribution with n independent trials, and probability p of the event of interest on each trial.

It may be observed that these probabilities are successive terms in the binomial expansion of $(p + (1 - p))^n$, so that obviously $\sum_{x=0}^{n} f(x) = 1$. Note that usually we write

$$q = 1 - p$$

It is straightforward to show that

$$\begin{aligned} E(X) &= np \\ Var(X) &= npq \end{aligned} \tag{2.25}$$

Suppose exactly nine people exit from the elevator. Let's first suppose that we know for sure that these nine people are completely unrelated to one another (why this assumption?). Our event of interest is the number of these specific nine who are from overseas. This number (X) could be 0, 1, 2, ..., 9. If we suppose p is the overall proportion of overseas visitors to the building, then the binomial distribution would be a good model for X. What is p? We suppose as before that we do not know p. So p here would be assigned a probability distribution which would model our uncertainty about p. This example illustrates important aspects of statistical inference. If our assumptions are correct, then the binomial distribution would be a good model of 'reality' – the actual number out of nine independent people who would exit the lift. This is derived from the theory of probability. Of course, you are free to assign whatever probability you like, but as argued in Chapter 1, to be useful in the world, probabilities should be assigned in agreement, and also in this case, any rational observer should agree that the binomial distribution applies to this situation, so should be a good model of empirical reality. However, if p is unknown, the information that it is a binomial distribution is not very useful by itself in order to be able to compute any actual probabilities. Statistical inference involves finding out something about the value of p from data. The first step is to construct, prior to the evidence, our probability distribution for p. If we really knew nothing, then the uniform distribution on [0, 1] would be an appropriate choice. Later, we will see how all this can be used for inference.

Now why did we have to assume that the nine people are completely unrelated to one another? Suppose that we did not have this assumption. Then, for example, clusters of people amongst the nine might be from the same families. So if one of these is overseas, all the family members are also likely to be overseas. This violates the assumption of independence, and so the binomial distribution would not apply. Similarly, in any estimation of p, the issue of independence would have to be taken into account: do we mean the proportion of all unique individual persons who enter the building, or do we mean the proportion of unique *groups* of people (like families, friends, business associates, etc.). Clear definitions of the variables and parameters involved are absolutely vital in statistical inference in order to avoid incorrect results.

The Beta Distribution

It was assumed in the binomial distribution discussion that the probability of event E occurring is known and equal to p. But suppose it is not known? As we have argued, in this case, p itself would have a probability distribution. If 'nothing' at all were known, the distribution of p could be modelled as *uniform*(0,1). But something might be known – for example, that p was closer to 1 than 0, or that its most probable value is 0.75, and so on. This can be modelled by the Beta distribution.

First, we define the function known as the Beta function, denoted by $B(a, b)$, for constants $a > 0$, $b > 0$:

$$B(a,b) = \int_0^1 u^{a-1}(1-u)^{b-1}\,du \tag{2.26}$$

Here a, b are any positive constants. For example, from the definition it follows that $B(1, 1) = 1$, $B(2, 1) = 0.5$, and so on. For example, in the case $B(1, 1)$:

$$B(1,1) = \int_0^1 du = 1$$

The Beta function has a close connection with another function, called the Gamma function. This is defined for $a > 0$ as

$$\Gamma(a) = \int_0^\infty u^{a-1} e^{-u} du \qquad (2.27)$$

It is easy to show that

$$\Gamma(a + 1) = a\Gamma(a)$$

and

$$\Gamma(1) = 1$$

It follows that if n in an integer, then

$$\Gamma(n + 1) = n(n - 1)\ (n - 2) \ldots 2.1 = n!$$

Hence, the Gamma function is a generalization of the idea of a factorial, since the Gamma function itself is defined for any positive a, and in the special case when a is an integer, we get the factorial.

Also, it is straightforward to show that

$$\Gamma(a)\Gamma(b) = \Gamma(a + b)\ B(a, b)$$

Therefore, we can rewrite the Beta function as

$$B(a,b) = \frac{\Gamma(a)\Gamma(b)}{\Gamma(a+b)}$$

Now we can use this to define the *Beta probability distribution*, for random variable X with domain [0, 1]:

$$f(x) = \frac{1}{B(a,b)} x^{a-1}(1-x)^{b-1},\ x \in [0,\ 1] \qquad (2.28)$$

The denominator $B(a, b)$ is necessary so that when we integrate $f(x)$ over the range [0, 1], we will obtain the correct answer, namely 1. We could equally well write the distribution as

$$f(x) \propto x^{a-1}(1-x)^{b-1},\ x \in [0,\ 1]$$

where the '*is proportional to*' symbol, $\propto$, includes the understanding that the constant of proportionality must be chosen to make the integral 1. We will often use this convention.

From the definitions of the mean (expected value) and variance, we can evaluate the integrals to find

$$E(X) = \frac{a}{a+b}$$

$$Var(X) = \frac{ab}{(a+b)^2 (a+b+1)} \tag{2.29}$$

The notation $X \sim$ Beta(a,b) reads 'X has a Beta distribution with parameters a and b'.

If in Equation (2.28), we set $a = 1$ and $b = 1$, then we have the *uniform*(0,1) distribution. Hence, the uniform distribution is a special case of the Beta.

Examples of the shape of the Beta distribution for various parameter values are shown in Figure 2.6. There are many possible different shapes for this distribution. Of course, when both parameters are 1, then the *uniform*(0,1) distribution is obtained as just mentioned. When the two parameters are equal, the distribution is symmetric about 0.5, which in this case is also the mean. When both parameters are less than 1, then the distribution is U-shaped. When one of the parameters is 1 and the other 2, then the distribution is a 45-degree line (not shown).

The Beta distribution is very important as a means for expressing our uncertainty about the value of a parameter that represents a proportion. It offers great flexibility through the appropriate choices of a and b.

The Negative Binomial Distribution

When there are Bernoulli trials, the binomial distribution is appropriate to model the number of occurrences of an event in n trials (fixed in advance). Instead, we can consider the *number of trials necessary to get r occurrences of the event*. For example, how long do you have to wait by the exit in the tallest building in Borgonia for a total of $r = 20$ overseas visitors to eventually leave (bearing in mind the discussion about independence above). Here the random variable X, denoting the number of trials awaited, has domain $X = r, r + 1, r + 2, \ldots$. In this case, X has no theoretical upper bound (we could wait forever).

One sequence corresponding to the event ($X = x$) is

$$\underbrace{111\ldots111}_{r-1}\underbrace{000\ldots000}_{x-r}1$$

The first $r - 1$ trials do result in the event. The next $x - r$ trials do not result in the event. Up to this point, the requirement for r occurrences of the event is not satisfied since there have been only $r - 1$ occurrences. This is, however, realized on the last event shown. For r occurrences of an event, the last one must always result in the event. This is the necessary rth occurrence.

This particular sequence has probability $(1 - p)^{x-r}p^r$. Since the last trial *must* be an occurrence of the event, there are $\binom{x-1}{r-1}$ such sequences that have exactly $r - 1$ ones and $x - r$ zeros (the very last one is always 1).

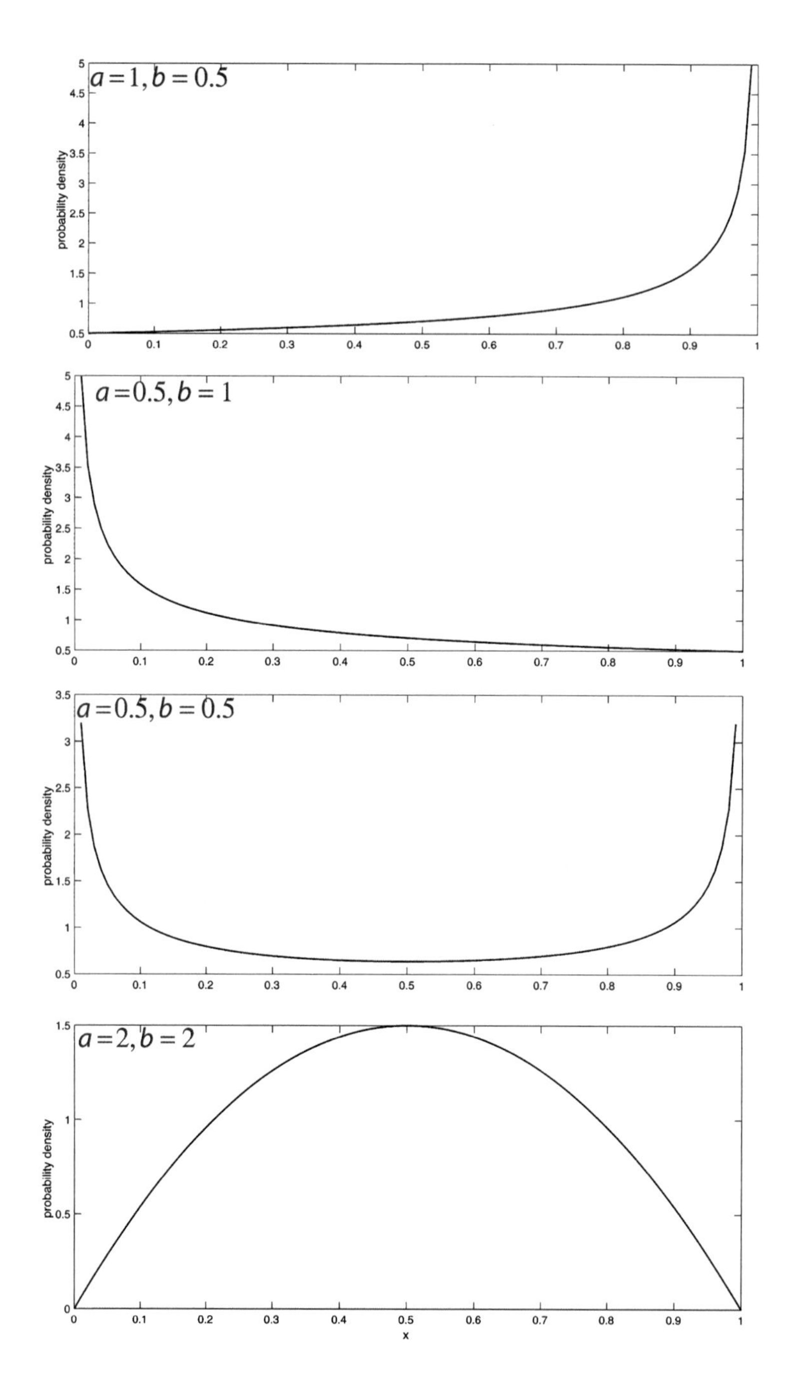

Figure 2.6

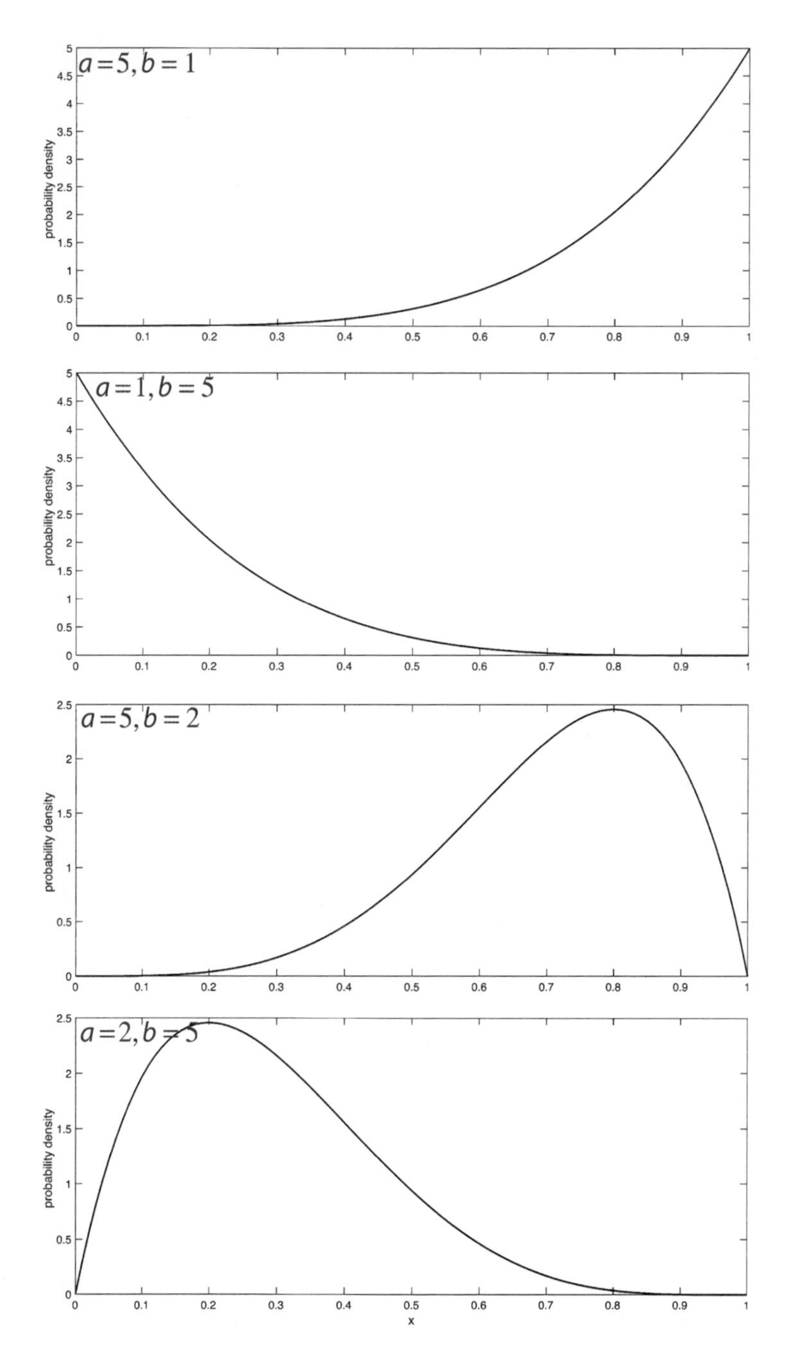

Figure 2.6 Examples of the shapes of the Beta distribution for various parameter values

Therefore, the pdf is

$$f(x)=\binom{x-1}{r-1}(1-p)^{x-r}p^{r},\ x=r,r+1,r+2,\ldots \tag{2.30}$$

Using the definitions,

$$E(X)=\frac{rq}{p}$$
$$Var(X)=\frac{rq}{p^{2}} \tag{2.31}$$

The Poisson Distribution

Suppose an event occurs at 'random' moments in time, and we are interested in modelling the probability of the number of occurrences in a certain time period. For the moment, suppose that it is a 'unit of time' in which we are interested, say the interval $[t, t+1]$ (meaning between time t and $t+1$ inclusive) for some fixed t. We suppose that whatever t the random process behaves the same, so that the specific value of t is immaterial. Now divide this time period up into n equal intervals, and let the probability of the event occurring in any time interval be p. We assume further that events across these equal time intervals are independent, and that the probability of two or more events in any of these intervals of length $1/n$ is negligible. If we let X be the random variable denoting the number of happenings, then $X \sim binomial(n, p)$. Now let n get larger and larger, and correspondingly p get smaller and smaller, but such that the mean rate of occurrence per unit time remains the same at μ. From Equation (2.25), we must have $\mu = np$. Let's examine the mathematics of this (you can skip this if you want). For the random variable X, we have the following distribution:

$$\binom{n}{x}p^{x}(1-p)^{n-x}=\frac{n(n-1)\ldots(n-x+1)}{x!}\left(\frac{\mu}{n}\right)^{x}\left(1-\frac{\mu}{n}\right)^{n-x}$$
$$=\left(\frac{\mu^{x}}{x!}\right)\left(\frac{n}{n}\right)\left(\frac{n-1}{n}\right)\ldots\left(\frac{n-x+1}{n}\right)\left(1-\frac{\mu}{n}\right)^{n-x}$$
$$=\left(\frac{\mu^{x}}{x!}\right)e^{-\mu},\ \text{as } n\to\infty$$

The $n\to\infty$ simply means that this expression holds for large n. Although the Poisson distribution has a meaning in its own right ('the number of events occurring at random in a given time period'), it is also an approximation to the binomial distribution under the conditions stated. The Poisson distribution is defined by

$$f(x)=\left(\frac{\mu^{x}}{x!}\right)e^{-\mu},\ x=0,1,2,\ldots \tag{2.32}$$

One important characteristic of the Poisson distribution which can be shown using Equations (2.13) and (2.16) (remembering to replace integrals by sums) is that the mean and variance are equal:

$$\begin{aligned} E(X) &= \mu \\ Var(X) &= \mu \end{aligned} \tag{2.33}$$

The restaurant at the top of the tallest building in Borgonia, The Bogo Tower Cafe Experience, prides itself on preparing the most creative and innovative cuisine in the entire region. Celebrities from around the world fight to dine there. When 2 years ago they opened a new wing, the entire following year's bookings were sold out in 2 hours after the reservation system went online. However, The Bogo Tower is also known locally as the Broken Tower (this is as close as possible to the true meaning, but it cannot be well translated from the original Borgonian). The reason is that the waiters, notwithstanding their professionalism and dedication, are famous for dropping trays of glasses, of course always brimming with the most expensive champagne.

The restaurant is open 365 days per year and 14 hours per day, and hence last year it was open for 5110 hours. (In leap years, it closes on February 29.) The number of dramatic glass breaking (and champagne wasting) events was 114. Well, this doesn't seem to be so many compared to the number of hours that the restaurant was open, but these stories have a life of their own.

The restaurant happens to be state owned, and Borgonian citizens are very proud of their extensive welfare state, which even extends to state ownership of extremely high-class restaurants. Some civil servants wondered whether the glass breakages were part of some organized campaign (or maybe champagne) against the state, or were truly random events. Borgonia operates a 5-day week following its ancient traditions, so that the restaurant is open for 70 hours per Borgonian week. So in a 365-day year, there were 73 Borgonian weeks.

For last year, the occurrence of breakages is given in Table 2.2.

Table 2.2 Frequency of glass breakages per week

Number of breakages per week	Frequency	Percentage	Theoretical percentage based on sample mean
0	21	28.8	21.0
1	18	24.7	32.8
2	17	23.3	25.6
3	8	11.0	13.3
4	7	9.6	5.1
5	2	2.7	1.6
≥5	0	0	0.6
Total	73	100	100

The sample mean and variance of the number of breakages (X) are

$$\bar{x} = 1.56,\ s^2 = 1.97$$

Recall that for the Poisson distribution, the mean and variance should be the same. Using the mean of 1.56, we can compute, using Equation (2.32), the probabilities of the number of breakages per unit time being 0, 1, 2, ... and compare with the actual results above. These theoretical percentages are shown in the last column of Table 2.2.

For example,

$$P(X = 2) = \frac{1.56^2}{2!}e^{-1.56} = \frac{2.4336}{2} \times 0.2101361 = 0.25$$

which is the probability. Multiply by 100 to transform it into a percentage, and we arrive at 25.6 as shown in Table 2.2.

Comparing the true results with the theoretical results, we can see that there is some similarity. At the moment, we don't have the tools to say how much the departure of the observed from the theoretical frequencies suggests an incompatibility of the breakages with the Poisson distribution. But at least, it seems possible that there is no conspiracy, that the breakages are 'more or less' what would be expected were they happening at random.

The Exponential Distribution

Let's stay with the same context as for the Poisson distribution. However, in this case, we're interested not in the number of events that occur but *the time between the events*. Call Y the random variable that denotes the time to the next event, and consider the probability that $Y \leq y$. The domain of this random variable is $[0, \infty]$. Again, consider that time is divided into intervals, each of size $1/n$, and that the mean rate of occurrence of the event is λ per unit time. Now in time y, there are ny intervals. In order for the time to the next event to occur in time less than or equal to y, the event must occur in the next time interval (the probability of this is p), or the one after that (probability is pq), or the time after that (probability is pq^2), and so on, where $q = 1 - p$. In other words, the event $Y \leq y$ is equivalent to occurrence in the next interval, or non-occurrence in the next interval followed by occurrence in the one after, and so on. Since these are exclusive events, we have for fixed intervals of size $1/n$

$$P(Y \leq y) = p\left(1 + q + q^2 + \ldots + q^{ny-1}\right) = p\left(\frac{1-q^n}{1-q}\right) = 1 - \left(1 - \frac{\lambda}{n}\right)^{ny}$$

As in the case of the Poisson distribution, consider what happens as $n \rightarrow \infty$. We use the fact that in general $\left(1 + \frac{x}{n}\right)^n \rightarrow e$ as $n \rightarrow \infty$.

Therefore,

$$P(Y \leq y) = 1 - e^{-\lambda y}$$

This is the *distribution function* $F(y)$. Recall that the pdf is the derivative of the distribution function, and therefore differentiating with respect to y to get the pdf,

$$f(y) = \lambda e^{-\lambda y},\ y \geq 0 \tag{2.34}$$

This is called the *exponential distribution*, and it is a good probability model for the 'time between random events'. It is possible to show using Equation (2.13) that

$$E(Y) = \frac{1}{\lambda}$$
$$Var(Y) = \frac{1}{\lambda^2} \tag{2.35}$$

Considering the set-up of the Poisson and exponential distributions together, we have an event that occurs at the rate λ per unit time, and hence obviously the mean (expected value) of a Poisson random variable is λ. On the other hand, the mean time *between* such events is $1/\lambda$. The mean number of people who enter the tallest building in Borgonia per hour, during working hours, is 30. Then the mean time between each successive person is 1/30 hours (2 minutes). It makes sense.

Now if the glass breakages in the Bogo restaurant were random, we would expect that the time between the breakages would follow the exponential distribution. Figure 2.7 shows the histogram of the times between successive breakages and a fitted exponential distribution. The mean time is 44.7 hours and the standard deviation is 52.2 (variance = 2725.4). Note that 70/1.56 = 44.8, where 1.56 was the mean of the corresponding Poisson distribution. This is what would be expected from Equations (2.34) and (2.35).

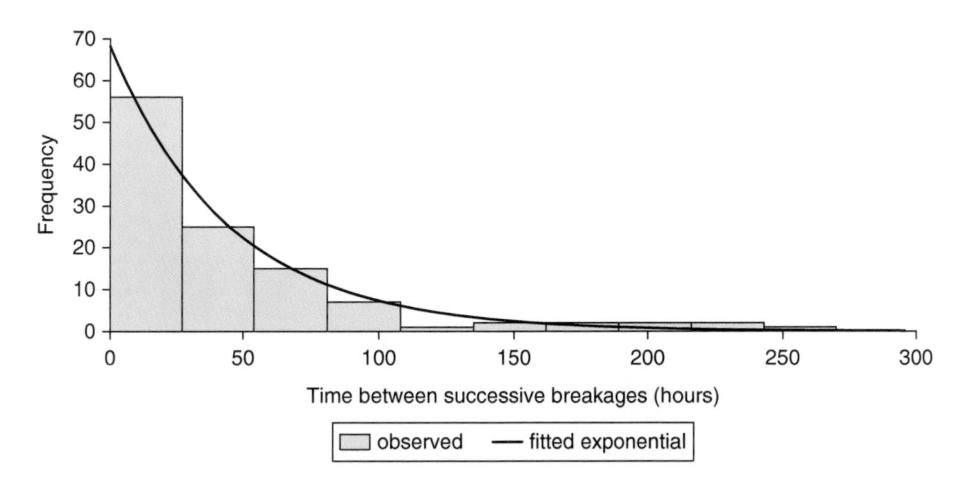

Figure 2.7 Time between successive glass breakages (hours)

The Gamma Distribution

The Gamma distribution is closely associated with the exponential as we shall see. The distribution is defined as

$$f(x) = \frac{\lambda^p}{\Gamma(p)} x^{p-1} e^{-\lambda x}, \ x \geq 0 \tag{2.36}$$

Comparing this with Equation (2.34), Equation (2.36) shows that the exponential distribution is a special case of the Gamma, with $p = 1$. The Gamma distribution can be derived as the distribution of the sum of p independent exponential variables, each with the same mean. In terms of the event occurring randomly in time, the Gamma distribution models 'the time between each successive p occurrences of the event'.

The parameter λ is called the rate and p is called the shape.

The mean and variance are

$$\begin{aligned} E(X) &= \frac{p}{\lambda} \\ Var(X) &= \frac{p}{\lambda^2} \end{aligned} \tag{2.37}$$

Compare this with Equation (2.35).

Going back to the breakages of glass in the restaurant, we consider the distribution of times between each three occurrences of the event. We would expect this to be a Gamma distribution. Figure 2.8 shows the histogram and the resulting fit. The mean is 48.0 and standard deviation is 50.8.

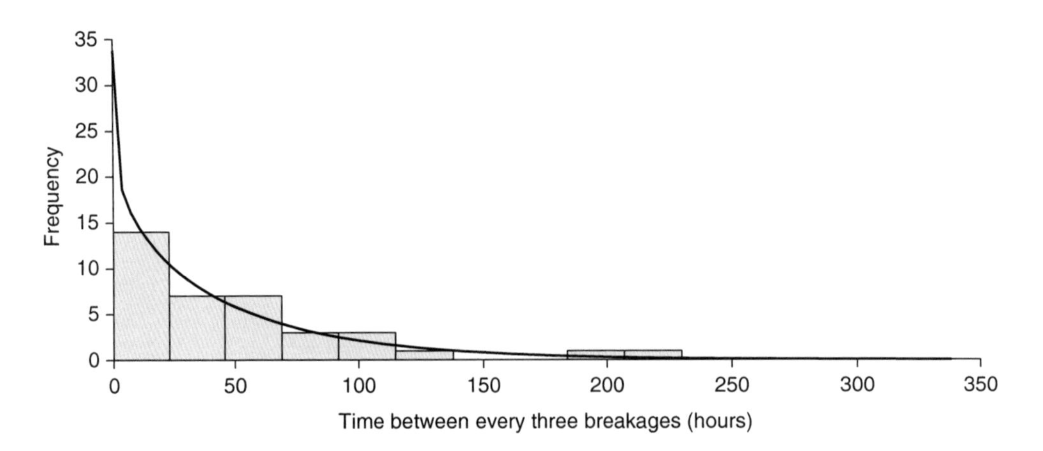

Figure 2.8 Time between each three successive glass breakages (hours)

Examples of the Gamma distribution are shown in Figure 2.9.

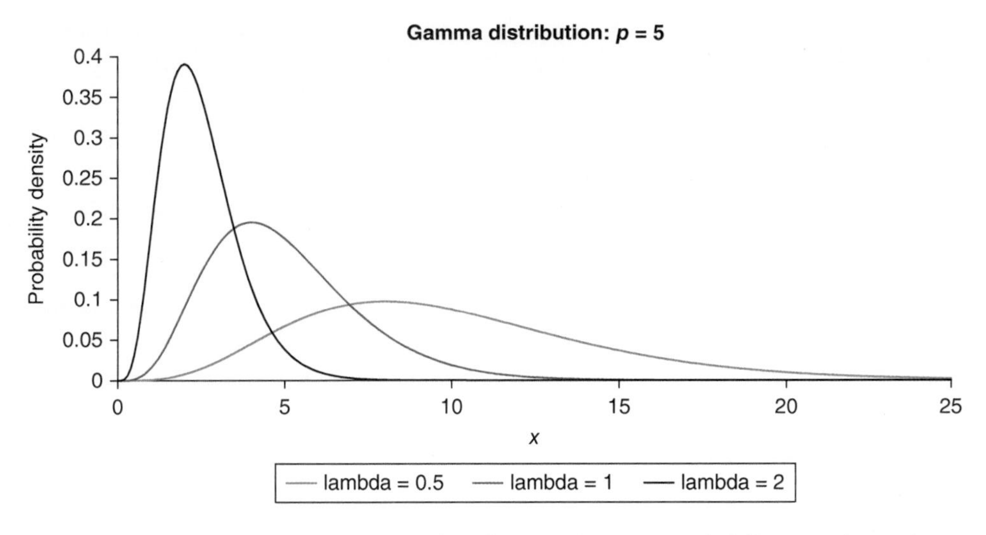

Figure 2.9 Examples of the Gamma distribution, for $p = 5$ and different values of λ

A special case of the Gamma distribution is called the chi-squared distribution, which turns up frequently in classical statistics. The chi-squared distribution with parameter n, the 'degrees of freedom', is a Gamma distribution with $p = n/2$ and $\lambda = 1/2$. In fact, the chi-squared distribution arises as the sum of squares of n independent standard normal distributions – which is the next topic.

The Normal Distribution

The normal (also called 'Gaussian') distribution is the most well-known distribution of statistics. It is defined as follows:

$$f(x) = \frac{1}{\sigma\sqrt{2\pi}} \exp\left(-\frac{1}{2}\left(\frac{x-\mu}{\sigma}\right)^2\right), -\infty < x < \infty \tag{2.38}$$

Moreover,

$$\begin{aligned} E(X) &= \mu \\ Var(X) &= \sigma^2 \end{aligned} \tag{2.39}$$

The distribution is completely specified by these two parameters. We will write this as $X \sim$ *normal*(μ, σ).

Suppose $X \sim$ *normal*(μ, σ), and consider the new random variable

$$Z = \frac{X-\mu}{\sigma} \tag{2.40}$$

It should be clear from Equation (2.39) that $Z \sim$ *normal*(0, 1). This is called *standardizing X* to make the mean 0 and the variance 1. The standard normal distribution is, therefore,

$$f(z) = \frac{1}{\sqrt{2\pi}} \exp\left(-\frac{1}{2}z^2\right), -\infty < z < \infty \tag{2.41}$$

The word 'normal' in the name of this distribution is not meant to imply that other distributions are abnormal. It is simply that the distribution reflects the fact that in many circumstances 'departures from the norm' happen in a symmetric, regular pattern. The distribution clusters around its mean value and tapers away symmetrically, and relatively rapidly, towards zero probability on each side of the mean. Approximately 68% of the distribution is within 1 standard deviation of the mean, 95% within 2 standard deviations, and 99% within 2.58 standard deviations. The distribution occurs frequently in nature, for example where the random variable in question is produced as an averaging effect of a large number of constituent 'causes'. Examples of the pdf of the normal distribution are shown in Figure 2.10, where we can see that as σ increases so the distribution becomes more spread out.

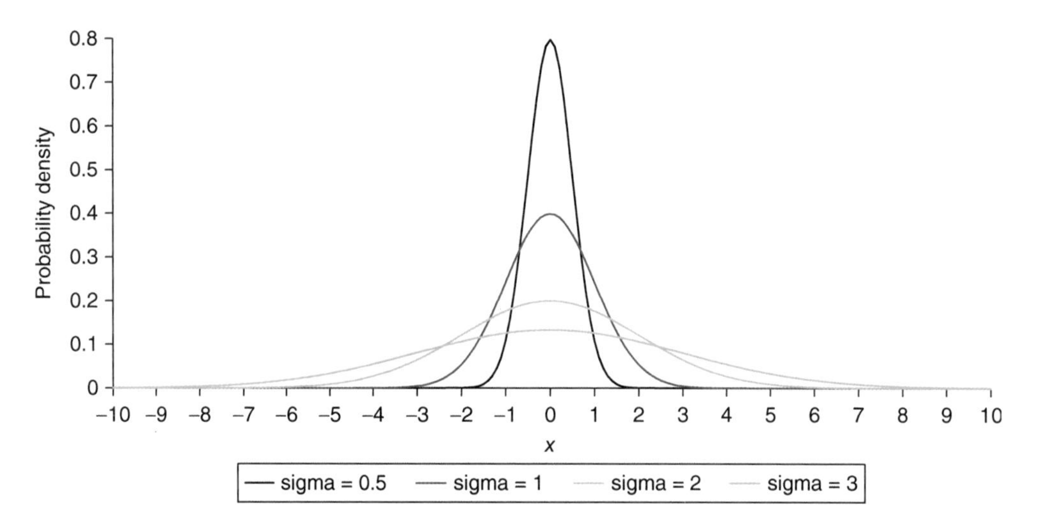

Figure 2.10 Examples of the normal distribution with mean 0 for various values of sigma

This idea that the distribution occurs naturally, through an averaging effect, is formalized in what is known as the central limit theorem (CLT). Suppose $X_1, X_2, \ldots, X_n$ are n independent and identically distributed (i.i.d.) random variables with finite mean μ and standard deviation σ. Then the mean of these variables is also a random variable:

$$\bar{X} = \frac{1}{n}\sum_{i=1}^{n} X_i$$

For 'large' n, the CLT is that

$$\bar{X} \sim normal\left(\mu, \frac{\sigma}{\sqrt{n}}\right)$$

In this case '~' should be read as 'approximately', and it is a result that holds with probability 1. It is an asymptotic result (i.e. holds for 'large' n). This is a powerful result, since the only requirement is that the n variables are i.i.d. and their common distribution can be anything (with finite mean and standard deviation), and this works. What is meant by 'large'? Around 30 is already enough.

Lognormal Distribution

The lognormal distribution, as its name suggests, is where the log of a random variable has a normal distribution. In other words, the random variable $Y > 0$ has a lognormal distribution when $\log(Y) \sim normal(\mu, \sigma)$. This is a skewed distribution (Figure 2.11), and we will see an example in the next chapter. The pdf is

$$f(y) = \frac{1}{y\sigma\sqrt{2\pi}} \exp\left(-\frac{1}{2}\left(\frac{\log(y)-\mu}{\sigma}\right)^2\right), y > 0 \tag{2.42}$$

$$E(Y) = \exp\left(\mu + \frac{\sigma^2}{2}\right)$$
$$Var(Y) = \exp(\sigma^2 - 1)\exp(2\mu + \sigma^2) \tag{2.43}$$
$$Median(Y) = \exp(\mu)$$

We give the median in this case, since we will use this in the next chapter.

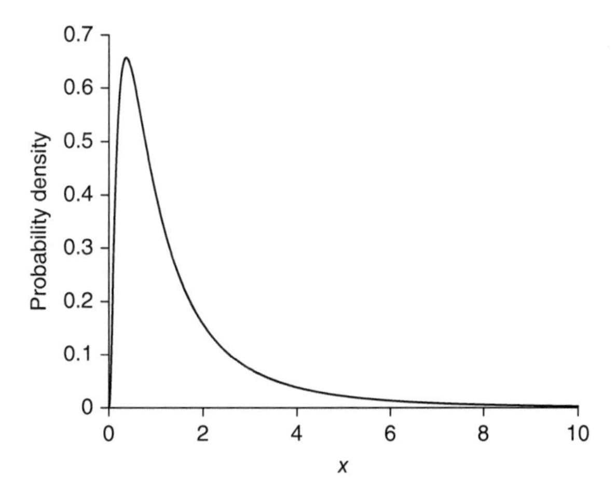

Figure 2.11 Lognormal distribution with parameters $\mu = 0$, $\sigma = 1$

The 'Student' *t* distribution

Another very frequently used distribution in statistics is called the Student *t* distribution. In this case, let Z be a *normal*(0,1) random variable, and let X be an independent chi-squared distribution with n degrees of freedom. Consider the ratio

$$t = \frac{Z}{\sqrt{X/n}}$$

This random variable t has the so-called Student t distribution with n degrees of freedom. It is called 'Student' since this was the pen name of William Gosset who presented this distribution in the early 20th century. This is a symmetric bell-shaped distribution around 0, and in fact, as n increases the distribution tends rapidly to normality. For n at around 30, the two distributions are very similar, so that the normal distribution may be used as an approximation to the t.

The pdf of the t distribution is as follows:

$$f(t) \propto \left(1 + \frac{t^2}{n}\right)^{-\left(\frac{n+1}{2}\right)}, -\infty < t < \infty \tag{2.44}$$

A special case of the t distribution occurs when $n = 1$ (1 degree of freedom). This is called the Cauchy distribution, and its pdf is

$$f(x) = \frac{1}{\pi\sigma\left(1 + \left(\frac{x-\theta}{\sigma}\right)^2\right)}, -\infty < x < \infty \tag{2.45}$$

The parameter θ provides the location of the distribution (where it is on the x-axis) and σ provides the scale (how spread out it is). However, θ and σ are *not* the mean and standard deviation, although θ is the median. In fact, this distribution has infinite mean and standard deviation (if you try to compute them, the integrals are infinite). The distribution is illustrated in Figure 2.12.

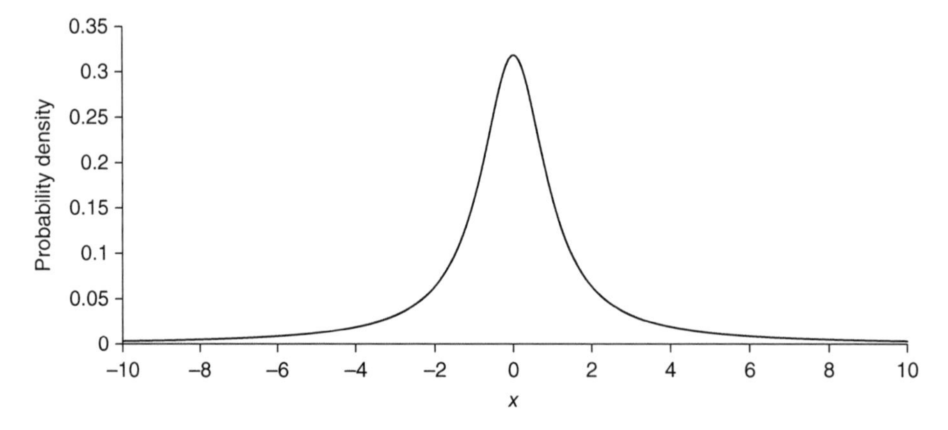

Figure 2.12 The Cauchy distribution with $\theta = 0$, $\sigma = 1$

Inverse Gamma Distribution

Suppose X has a Gamma distribution; then the distribution of the reciprocal $Y = 1/X$ has an inverse Gamma distribution. This will be used in the next chapter. The pdf is

$$f(y) \propto y^{-\alpha-1} \exp\left(-\frac{\beta}{y}\right), y > 0 \tag{2.46}$$

where the shape parameter $\alpha > 0$ and scale parameter $\beta > 0$.

$$\begin{aligned} E(Y) &= \frac{\beta}{\alpha - 1}, \quad \alpha > 1 \\ Var(Y) &= \frac{\beta^2}{(\alpha - 1)^2 (\alpha - 2)}, \quad \alpha > 2 \end{aligned} \tag{2.47}$$

The Logistic Distribution

The logistic distribution has a special role in the analysis of ordinal data – for example, responses to questionnaires such as 'How much do you like ...?' with possible answers ranging across the integers only, for example from 1 to 7 with meaning 'Not at all' (score 1), ..., 'Neither like nor dislike' (score 4), ..., 'Very much' (score 7). We will discuss this application in detail in Chapter 6.

The cumulative distribution function is

$$F(y) = \frac{1}{1 + e^{-\left(\frac{y-\mu}{\sigma}\right)}}, -\infty < y < \infty$$

The pdf is

$$f(y) = \frac{e^{-\left(\frac{y-\mu}{\sigma}\right)}}{\sigma\left(1 + e^{-\left(\frac{y-\mu}{\sigma}\right)}\right)^2}$$

The mean and variance are

$$E(Y) = \mu$$

$$Var(Y) = \frac{\sigma^2 \pi^2}{3}$$

Figure 2.13 shows the logistic distribution with $\mu = 0$ and $\sigma = 1$.

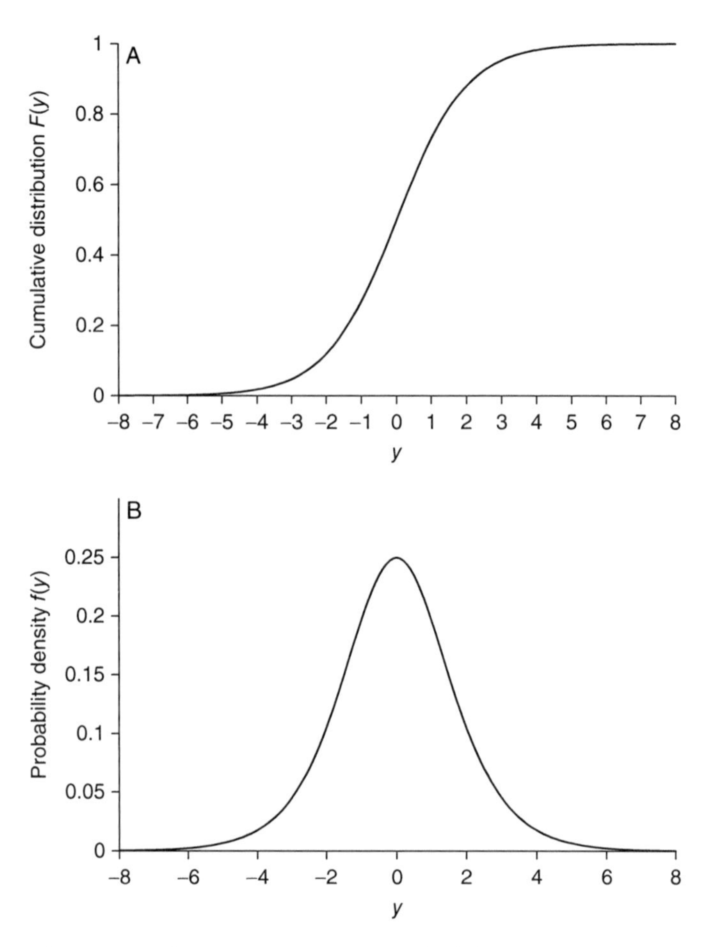

Figure 2.13 The logistic distribution with mean 0: (A) cumulative distribution function and (B) probability density function

Multivariate Distributions

There now follows a discussion that extends our coverage to distributions that simultaneously involve the joint distributions of two or more variables – namely, multivariate distributions. Along with convenient ways of describing these distributions, we will be able to conclude this chapter with consideration of some important distributions in statistics.

Distribution and Density Functions

Up to now, we have been treating single random variables and looking at a range of distributions. However, most problems in statistics involve the simultaneous consideration of many random variables. For example, if we consider a sample of n people in Borgonia and consider the total wealth of each (however measured), we have n random variables, one for each of the n people. Until we collect the data and have the actual values of these variables, they are considered as random variables. Moreover, if some of the n people are somehow connected to or related to one another, there will be relationships between these variables.

If X and Y are two *discrete* random variables, then the event $(X \le x) \cap (Y \le y)$ has probability

$$P((X \le x) \cap (Y \le y)) = F(x, y) \text{ (the cumulative distribution function)}$$

and the event $(X = x) \cap (Y = y)$ (X takes the value x and Y takes the value y) has probability

$$P(X = x \cap Y = y) = f(x, y) \text{ (the pdf)} \tag{2.48}$$

Another notation for the joint probability is simply $P(X = x, Y = y)$, where the comma is also read as 'and'.

Similarly if X and Y are continuous, then $f(x, y)dxdy$ is the probability that an observation on (X, Y) will be in a small neighbourhood around (x, y). This defines the joint probability distribution:

$$P(X < x, Y < y) = \int_{-\infty}^{x} \int_{-\infty}^{y} f(u, v)dudv = F(x,y) \tag{2.49}$$

As before, $F(x, y)$ is the distribution function and $f(x, y)$ the pdf, with exactly the same meanings as in the univariate case. The integral over the whole domain of the random variables is equal to 1, since this is the probability of the 'certain event' (the random variables take some value in their range).

This is extended in an obvious way to the joint distribution of several variables $(X_1, X_2, \ldots, X_n)$, which is denoted by $f(x_1, x_2, \ldots, x_n)$.

Given a joint distribution of a set of random variables, the distribution of any subset of this set can be found by integrating out all the others. These are then called the *marginal distributions*. For example, from $f(x, y)$, we could find the distribution of X alone by integrating out Y. (Of course, replace 'integrate' by 'sum' in the discrete case.) Hence, if $f(x, y)$ is the joint distribution of (X, Y) then the distribution of X alone is

$$\int_{-\infty}^{\infty} f(x, y)dy \tag{2.50}$$

and similarly integrate out X to get the distribution of Y.

We can define conditional distributions and independence directly from the axioms A8 and A9 of Chapter 1. The joint distribution of any two variables is related to the conditional distribution by

$$f(x, y) = f(y|x)\, f(x) \tag{2.51}$$

This extends to any number of variables – for example,

$$f(x_1, x_2, x_3) = f(x_1, x_2|x_3)\, f(x_3) = f(x_1|x_2, x_3)\, f(x_2|x_3)\, f(x_3) \tag{2.52}$$

The random variables are *independent* if and only if the joint distribution of all of them and each subset of 'marginal distributions' factorizes into the product of the appropriate set of (marginal) univariate distributions. Hence, $X_1, X_2, \ldots, X_n$ are independent if and only if

$$f(x_1, x_2, \ldots, x_n) = f(x_1)\, f(x_2) \ldots f(x_n) \tag{2.53}$$

and that this factorization holds for every possible subset of $X_1, X_2, \ldots, X_n$.

Summary Measures and Relationships

Just as in the case of single variables, where we can summarize their distributions with measures such as mean and standard deviation, so we can do the same for multivariate distributions. The notions of expected value and variance directly carry over. For example, for the joint distribution $f(x, y)$

$$\begin{aligned} E(X) &= \int_{-\infty}^{\infty} x\left[\int_{-\infty}^{\infty} f(x,y)\,dy\right]dx \\ &= \int_{-\infty}^{\infty}\int_{-\infty}^{\infty} xf(x,y)\,dy\,dx \end{aligned}$$

The inner integration finds the marginal distribution for X by integrating out y, and then we just use Equation (2.13). We often write $E(X)$ as μ_X or μ_x.

Similarly,

$$Var(X) = \int_{-\infty}^{\infty}\int_{-\infty}^{\infty} (x-\mu_x)^2 f(x,y)\,dy\,dx \tag{2.54}$$

and similarly for Y, with obvious extension to multiple variables.

Whenever there is more than one variable, the question of *relationship* between the variables immediately arises. One measure of this is the so-called *covariance* between two variables, and a normalized version of it called the *correlation*. The covariance between two variables is defined as

$$\begin{aligned} Cov(X, Y) &= E[(x - \mu_x)(y - \mu_y)] \\ &= E(XY) - \mu_x\mu_y \end{aligned} \tag{2.55}$$

This measures the degree of *linear relationship* between X and Y. Note that

$$Cov(X, X) = Var(X) \tag{2.56}$$

Also when X and Y are independent, $E(XY) = E(X)E(Y)$, and therefore, $Cov(X, Y) = 0$.

The correlation coefficient is the covariance normalized by the standard deviations of the two variables:

$$\rho(X,Y) = \frac{Cov(X,Y)}{\sqrt{Var(X)Var(Y)}} \tag{2.57}$$

The correlation coefficient is always between –1 and +1. It is 1 when there is an *exact* positive linear relationship between X and Y and –1 when there is an exact negative linear relationship. Suppose $Y = \alpha + \beta X$ for fixed constants α and β. Then by substituting the expression for Y in Equation (2.57), you should be able to show that $\rho(X, Y) = \pm 1$; it is 1 if $\beta > 0$ and –1 if $\beta < 0$.

Models that involve approximate linear relationships between variables are the mainstay of statistical analysis and will be coming up again and again in subsequent chapters.

Note that when we have several random variables, $\mathbf{X} = (X_1, X_2, \ldots, X_n)$, as well as the expected value and variance of each one, there is also the covariance matrix, which is a matrix consisting of all the covariances and with the variances down the main diagonal.

$$Cov(\mathbf{X}) = \begin{pmatrix} \sigma_1^2 & \sigma_{12} & \cdots & \sigma_{1n} \\ \sigma_{21} & \sigma_{22}^2 & \cdots & \sigma_{2n} \\ \vdots & \vdots & \ddots & \vdots \\ \sigma_{n1} & \sigma_{n2} & \cdots & \sigma_n^2 \end{pmatrix} = \Sigma \tag{2.58}$$

Here the entries are $\sigma_{ij} = Cov(X_i, X_j)$. Note that this is a symmetric matrix, since $Cov(X_i, X_j) = Cov(X_j, X_i)$. By dividing each entry by the product of the corresponding standard deviations from Equation (2.57), we obtain the correlation matrix:

$$Corr(X) = \begin{pmatrix} 1 & \rho_{12} & \cdots & \rho_{1n} \\ \rho_{21} & 1 & \cdots & \rho_{2n} \\ \vdots & \vdots & \ddots & \vdots \\ \rho_{n1} & \rho_{n2} & \cdots & 1 \end{pmatrix} \equiv \boldsymbol{P}$$

where $\rho_{ij} = \rho_{ji}$ is the correlation between X_i and X_j.

A Probability Distribution Over Correlation Matrices

A probability distribution that will be useful later is one that has its domain over the class of *correlation matrices* $\boldsymbol{P}$. This is known as the LKJ distribution. It is named after Lewandowski, Kurowicka, and Joe (2009). It can be defined as

$$f(\boldsymbol{P}|\eta) \propto |\boldsymbol{P}|^{\eta-1}$$

where $|.|$ refers to the determinant of the matrix and $\eta > 0$ is a parameter. Let's consider an example of a 2×2 correlation matrix:

$$\boldsymbol{P} = \begin{pmatrix} 1 & \rho \\ \rho & 1 \end{pmatrix}$$

The determinant $|\boldsymbol{P}| = (1-\rho^2)$. Hence, in this case the density function is

$$f(\boldsymbol{P}|\eta) \propto (1-\rho^2)^{\eta-1}$$

This is somewhat like the Beta distribution. When $\eta = 1$, it is the uniform distribution. When $\eta > 1$, then the maximum density occurs when $\rho = 0$, in which case, the correlation matrix $\boldsymbol{P}$

is the identity matrix. When $\eta < 1$, then the maximum density occurs towards the boundaries $\rho = \pm 1$. These observations apply without change to higher dimensions of $\boldsymbol{P}$.

The Multivariate Normal Distribution

A generalization of the normal distribution to multiple variables gives rise to the *multivariate normal distribution*. The pdf of the joint distribution of $\mathbf{X} = (X_1, X_2, ..., X_n)$ is

$$f(x_1, x_2, \ldots, x_n \mid \mu, \Sigma) \propto \exp\left(-\frac{1}{2}(\boldsymbol{x} - \mu)^T \Sigma^{-1} (\boldsymbol{x} - \mu)\right) \tag{2.59}$$

Here $\boldsymbol{x}$ is a column vector of the random variables $(x_1, x_2, ..., x_n)$,

$$\boldsymbol{x} = \begin{bmatrix} x_1 \\ x_2 \\ \vdots \\ x_n \end{bmatrix}$$

and $\boldsymbol{\mu}$ is a column vector of the corresponding expected values and $\boldsymbol{\Sigma}$ is the covariance matrix. The superscript T refers to *transpose*, where the rows become columns and the columns rows, so that, for example,

$$\boldsymbol{x}^T = [x_1, x_2, \ldots, x_n]$$

This distribution looks like a bell in the case of two variables, and in the general case, all of the marginal distributions are individually normal.

Consider what happens if all the covariances are 0. Then $\boldsymbol{\Sigma}$ is a diagonal matrix, with the variances on the main diagonal, and 0 elsewhere. Therefore, $\boldsymbol{\Sigma}^{-1}$ has the inverses of the variances on the main diagonal. In this case, the exponent is

$$-\frac{1}{2}(\boldsymbol{x} - \mu)^T \Sigma^{-1} (\boldsymbol{x} - \mu) = -\frac{1}{2}\sum_{i=1}^{n}\left(\frac{x - \mu_i}{\sigma_i}\right)^2$$

so that the joint distribution factorizes into the product of $normal(\mu_i, \sigma_i^2)$ distributions. By Equation (2.53) this would mean that the random variables are independent. The normal distribution is the only distribution that has the special property that when the covariances (or correlations) are all 0 then the random variables are independent.

The Dirichlet Distribution

The Dirichlet distribution is a generalization of the Beta distribution to the multivariate situation. $\boldsymbol{X}$ is a vector of n non-negative random variables, all between 0 and 1, with their sum equal to 1. The density function for the Dirichlet distribution is defined as

$$f(x_1, x_2, \ldots, x_n \mid \alpha) \propto x_1^{\alpha_1 - 1} x_2^{\alpha_2 - 1} \ldots x_n^{\alpha_n - 1} \tag{2.60}$$

where $x_n = 1 - x_1 - x_2 - \ldots - x_{n-1}$, $x_i \geq 0$, and all $\alpha_i > 0$.

Compare this with Equation (2.28), and it can be seen that the Beta distribution is a special case. Moreover the marginal distribution of any subset of the variables has a Dirichlet distribution. The univariate marginal distributions are Beta. Notice also that because of the constraint imposed by the sum of the variables always being 1, they are not independent, since any one of the variables may be expressed as one minus the sum of the others. Hence, the joint distribution cannot factorize into the product of the marginals.

The Dirichlet distribution is appropriate for modelling the joint distribution of random variables such as a set of proportions that sum to 1. For example, the visitors to the tallest building in Borgonia might be from Asia, Africa, North America, South America, Antarctica, Europe, or Australasia. Let θ_i, $i = 1, \ldots, 7$, be the proportions of visitors in a year from each of these continents, respectively. Then the probability distribution of the proportions of visitors from these continents might be modelled as Dirichlet.

The Multinomial Distribution

This distribution stands in the same relationship to the Dirichlet distribution as the binomial distribution to the Beta distribution. It is a generalization of the binomial distribution. The binomial distribution arises from a sequence of independent trials each of which has two outcomes – 'Event occurs' or 'Event does not occur'. The multinomial distribution has the same set-up of independent trials, but each trial results in one of k outcomes, 1, 2, ..., k. Suppose that the probability of the ith outcome is p_i on each trial. Then,

$$\sum_{i=1}^{k} p_i = 1$$

Suppose that there are n trials and X_i denotes the number of times that outcome i occurs. Then following the same reasoning as for the binomial distribution

$$P(X_1 = x_1, X_1 = x_2, \ldots, X_k = x_k) = f(x_1, x_2, \ldots, x_k) = \frac{n!}{x_1! x_2! \ldots x_k!} p_1^{x_1} p_2^{x_2} \ldots p_k^{x_k}$$

where all $x_i \geq 0$,

$$\sum_{i=1}^{k} x_i = n$$

and

$$p_k = 1 - p_1 - p_2 - \ldots - p_{k-1}$$

In the case $k = 2$, this reduces to the binomial distribution.

Summary

This chapter has introduced the idea of probability distributions over random variables. A probability distribution may be represented as a distribution function or as a density function. Summary measures such as the mean and variance (or standard deviation) were defined. We introduced a number of distributions that are useful either for modelling aspects of physical reality or to express uncertainty about an unobservable parameter.

We introduced the notion of multivariate distributions and defined the idea of conditional distributions and independence. These follow naturally from the axioms of probability introduced in Chapter 1.

Online Resources

The R examples that go along with this chapter can be found at

www.kaggle.com/melslater/slater-bayesian-statistics-2

All of the R code can be interactively executed. Parameter values for the various distributions can be changed in order to see the effects.

Three

MODELS AND INFERENCE

Introduction

The people of Borgonia believe that they live in a particularly egalitarian society, where there are no gross inequalities in income and standard of living across the population. Moreover, they believe that their society does not suffer discrimination on the basis of sex, gender, age, race, or religion. For example, they claim that the proportions of women scientists, engineers, and doctors are equal to men at every level rather than men being disproportionally high at the higher levels of the professions. How might we verify that?

Consider an example. In the UK, the proportion of all staff who work in the National Health Service (NHS) who are women is 77%. The proportion of female staff who are doctors or dentists is 5%, and the proportion of male staff who are doctors or dentists is 22%. These results are from the statistics of the NHS and are population values – in other words, these statistics are the 'true' values of these parameters.[1] Now suppose you have an appointment with a doctor or dentist. What is the probability that the doctor or dentist will be a woman? For the sake of illustration, we can suppose that the doctor or dentist you are going to meet has been selected at random from all doctors and dentists – or at least that the various distributions are uniform – independent of location (this is an unlikely assumption).

Let's use the symbol D for doctor or dentist, W for woman, and M for man. Then, we are interested in the probability $P(W \mid D)$. Given that you are going to meet a D, what is the probability of that person being female?

From A8 of Chapter 1:

$$P(W|D) = \frac{P(D|W)P(W)}{P(D)}$$

$$P(D|W) = 0.05$$

$$P(W) = 0.77$$

So we have the numerator of the expression as 0.0385.

We can rewrite the denominator as

$$P(D) = P(D|W)P(W) + P(D|M)P(M)$$

All this says is that a doctor is either M or W: $(D \cap M) \cup (D \cap W)$. Someone who is a doctor is either a doctor and a man, or a doctor and a woman (for simplicity, leaving aside other gender roles). But A8 gives us that

$$P(D \cap M) = P(D|M)P(M)$$

$$P(D \cap W) = P(D|W)P(W)$$

[1]www.nhsemployers.org/~/media/Employers/Publications/Gender%20in%20the%20NHS.PDF

Moreover, these two events are exclusive (a doctor cannot be a man and a woman – again for the sake of simplicity, we are leaving aside other possibilities in this discussion). Hence from Chapter 1, A7, the probability of 'doctor and man' or 'doctor and woman' is the sum of the probabilities.

Hence, putting all this together, we have

$$\begin{aligned} P(W|D) &= \frac{P(D|W)P(W)}{P(D|W)P(W)+P(D|M)P(M)} \\ &= \frac{0.0385}{0.0385+(0.22\times 0.23)} \\ &= 0.432 \end{aligned} \tag{3.1}$$

Following the same reasoning we could also find that

$$\begin{aligned} P(M|D) &= \frac{P(D|M)P(M)}{P(D|W)P(W)+P(D|M)P(M)} \\ &= \frac{0.22\times 0.23}{(0.77\times 0.05)+(0.22\times 0.23)} \\ &= 0.568 \end{aligned} \tag{3.2}$$

Note that the two probabilities sum to 1, as they must do since the doctor will be M or W in this discussion. This is always the case, as we can see by summing Equations (3.1) and (3.2).

Equations (3.1) and (3.2) are examples of Bayes' theorem. Another way in which Equation (3.1) could be written is

$$P(W|D) \propto P(D|W)P(W) \tag{3.3}$$

where $\propto$ means 'is proportional to'.

If we did not know the denominator $P(D)$, then from the numerators of Equations (3.1) and (3.2), since we know that the sum of these probabilities must be 1, we can always find the denominator.

The probability that a doctor selected at random would be a woman might seem to be unexpectedly high. This is because of the confusion between two quite different statements: 'the proportion of women who are doctors' (5%) and 'the proportion of doctors who are women' (now we know it is 43%). Although the proportion of women who are doctors is very low, the proportion of women in the NHS is relatively high. Here, probability theory has given us an answer that is somewhat counter to intuition (which is good, because if we could always rely on intuition, we would not need mathematics or science).

Why is Equation (3.1), which denotes Bayes' theorem, so important? Let's give different meanings to W, M, and D. Suppose D stands for *observable **d**ata*. Suppose W refers to some hypothesis (a statement about the ***w**orld*). W is *unobservable* (at least right now): until the doctor arrived, you did not know if the person would be male or female. At the time that you made the probability assessment, the outcome itself was unobservable. Another way to say this is that you had the *hypothesis* 'The doctor who will show up will be a woman'.

Considering Equation (3.3), $P(D|W)$ is called the likelihood. It is the probability of observing the particular data D conditional on the hypothesis W being true. In fact, anything that is proportional to this probability is the likelihood. Typically, it is possible to compute the likelihood based on probability theory, given the situation. In the particular case of our example with the doctors, this probability was given from population statistics. $P(W)$ is called the prior probability. It is the probability of W prior to observing the data. In our particular case, this was known from population statistics. However, we are usually not so lucky, especially if W is a hypothesis whose truth or falsity cannot be directly observed in the world. Normally, $P(W)$ is a subjective probability assigned on the basis of prior knowledge, or if there really is no prior knowledge, then we would have to use a prior probability that reflected this. For example, if we had no idea whatsoever about the relative proportions of males and females who work in the NHS, then a natural choice would be $P(W) = 0.5$.

$P(W|D)$ is usually called the posterior probability of W – posterior because it is after actually observing the data.

Putting this together and using Equation (3.3), we can write as follows:

$$\text{Posterior probability} \propto \text{likelihood} \times \text{prior probability}$$

Bayes' theorem can therefore be thought of as a way to update your prior probability in the light of data.

In Borgonia, we have the following situation. The proportion of women amongst all staff in the Borgonia Health Service (BHS) is 15%:

$$P(W) = 0.15$$

The proportion of women who are doctors is 85%:

$$P(D|W) = 0.85$$

The proportion of men who are doctors is 45%:

$$P(D|M) = 0.45$$

You are going to meet a doctor. What is the probability that the doctor you meet is a woman?

$$P(D|W)P(W) = 0.85 \times 0.15 = 0.1275$$

$$P(D|M)P(M) = 0.45 \times 0.85 = 0.3825$$

$$P(W|D) = \frac{0.1275}{0.1275 + 0.3825} = 0.25$$

Therefore, the probability that the doctor you meet is a woman is 25%. Perhaps Borgonia is not so egalitarian as people believe. This result, as the one above, might not be expected from the raw figure that 85% of female staff are doctors. As we commented above, if results always followed common sense, then there would be no need for probability theory.

Inferences About Parameters

Let's return to the issue of equality, and consider income distribution in the population of Borgonia. How can we verify whether it is indeed the case that there is a more egalitarian distribution compared to other countries? Given sufficient resources, we could carry out a survey of the working population over a well-specified time period. We would have to get all the definitions watertight – for example, is the unit of analysis an individual or a family? If a family, what constitutes a family? What do we mean exactly by the 'working population'? What do we mean by income? Is it just salaries or other sources of income such as interest or shareholdings? Is it liquid income (e.g. actual money), or does it include the value of property and other goods, and so on? Well, let's suppose all these issues are sorted out, and the focus is on individuals. There are millions of them, and we are not going to be able to obtain data on all of them. (Although such data exist in the Borgonia Tax Registry, very strict data protection laws allow no access.) So we decide on a sample of 1000 people, and the idea is to ask them to provide, under strict rules of privacy, their last tax return. (This assumes, of course, that in Borgonia there is little tax evasion.)

How that sample is chosen could be the subject of another book (or two). You might think, 'Oh my friend works for a company where there are more than 1000 employees.' We can ask the first 1000 employees of that company. This would not be useful, since the distribution of income in that company might be far from representative of the society as a whole. Alternatively, we could send a text message to mobile phone users until we had signed up 1000, or use social media. Again, this wouldn't work at all, since mobile phone and social media users might form a particular stratum of society, and be quite unrepresentative.

To avoid biases, the best way would be to have some kind of list of all working people in Borgonia and select a random sample from that. A random sample? To be truly random would be very difficult – perhaps involving some radioactive decay machine. Typically, a random sample would be drawn using a pseudo-random number generator available from a computer program. It is called 'pseudo' because it generates sequences of numbers that apparently have the property of being random, but, in fact, they are generated by a formula and so cannot be 'really' random.

Other methods of drawing samples are worth learning about but are beyond the scope of this book. For example, if we knew various properties of the population such as proportions of men and women, age distribution, distribution of workers in different sectors of industry, geographical location, and so on, then we would choose a *stratified* sample – where we would sample randomly within each of the sectors identified, and in accordance with their proportions in the population. However, for the sake of simplicity and sticking to the focus of this book, we will assume that the sample is *simple random*. This means essentially having a single list of all the population and using a random sampling procedure to select individuals. Again to keep things simple, we suppose that we collect on each of the 1000 people their annual after-tax income in b€ (Borgonian euros).

At the end of this lengthy and expensive process, we will have 1000 numbers. Just looking at those numbers is not going to be very informative – there is too much information. You could say,

'Ok, just do something manageable and look at the first 100 numbers.' If that is the method to be adopted, why bother sampling 1000? We could have saved a lot of effort in just choosing 100.

We have to summarize these data in order to obtain some understanding. A summary is in essence a model of the data as a whole. Before considering this in more detail, we consider one particular, but very important, aspect of these data that pertains to the issue of the egalitarian nature of Borgonian society – the proportion of people with income below the poverty line. Then, we will return to summaries of the distribution of income overall.

Inference About a Proportion

So for the moment, instead of tackling overall income distribution, let's consider the proportion of people below the 'poverty line'. Now the poverty line is not absolute but varies over time, and from country to country. But during the particular period of interest, the poverty line is defined as *P* b€. So any individual with an income less than *P* b€ is, by definition, below the poverty line. Let's let θ be the proportion in the population with an income below the poverty line. For example, in the UK, the proportion of households below this line is around 0.2, pre-pandemic and at the time of writing. Is Borgonia any better?

We have the sample of n individuals, and for each individual ($i = 1, 2, \ldots, n = 1000$), we denote their income as y_i, and it is simple to check whether $y_i < P$. If so, then that individual is below the poverty line, otherwise not. Let X be a random variable denoting the number of people out of n who are below the poverty line. Let the true proportion in the whole population be denoted by θ. This situation exactly corresponds to the binomial distribution discussed in Chapter 2. Hence, we can write using Equation (2.24)

$$P(X = x|\theta) = f(x|\theta) = \binom{n}{x}\theta^x(1-\theta)^{n-x}, x = 0, 1, 2, \ldots, n$$

Here $f(x|\theta)$ is the probability that x individuals are below the poverty line conditional on knowing θ, meaning that if we knew θ, we could calculate this probability for any x. However, θ is unobservable. Using Bayes' theorem, we can also write

$$g(\theta|x) \propto f(x|\theta)p(\theta)$$

Here, $g(\theta|x)$ is the probability density function for θ conditional on x, and $p(\theta)$ is the prior distribution for θ. If we knew the prior distribution for θ, we could actually find the posterior distribution, and then be able to make probability statements about this proportion. However, recall that θ is unobservable. We cannot ever 'know' the prior distribution, we can only choose one that reflects our prior information about the parameter of interest. So what should we choose?

Unfortunately, there are no published Borgonian official statistics about the proportion of people below the poverty line, and in fact, if there were, we would be wasting our time, because we would already know the answer to our question. So we are in a 'know nothing' situation. Which prior distribution is appropriate for that? In Chapter 2, we introduced the (continuous) uniform distribution which assigns equal probabilities over a continuous range. So we let the prior distribution be $\theta \sim uniform(0, 1)$ since the proportion obviously has to be in the range 0 to 1.

From Chapter 2, we know that this is a particularly easy distribution: that is, $p(\theta) = 1$ for θ in the range 0 to 1 and 0 outside this range. Therefore, the posterior distribution is

$$g(\theta|x) \propto \binom{n}{x} \theta^x (1-\theta)^{n-x}$$

It is important to understand that here θ is the random variable, and x is a known value, since we know it from our sample. For θ in the range 0 to 1 (and 0 otherwise, which from now on we can ignore), this is a probability density function for θ, and n and x are known values. Hence, we can simplify the expression even further to

$$g(\theta|x) \propto \theta^x (1 - \theta)^{n-x}$$

since $\binom{n}{x}$ is a constant and does not involve θ.

Now compare this with the Beta distribution introduced in Chapter 2 (Equation 2.26). We can see that the posterior distribution must be

$$\theta|x \sim Beta(x+1, n-x+1)$$

From this, we can compute the mean and variance of the distribution following Equation (2.29):

$$E(\theta|x) = \frac{x+1}{n+2}$$
$$Var(\theta|x) = \frac{(x+1)(n-x+1)}{(n+2)^2(n+3)} \tag{3.4}$$

Figure 3.1 shows the actual income distribution from our sample. (Note that the *x*-axis is in units of 10,000s.)

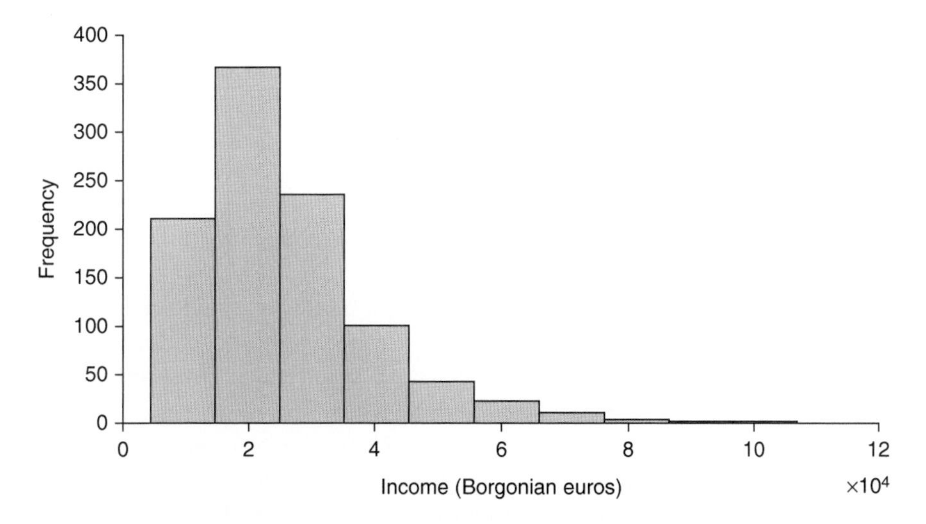

Figure 3.1 Income distribution of 1000 individuals selected at random from the Borgonian working population

The legal definition of 'below the poverty line' is an income less than 20,000 b€. The number of individuals in the sample with an income less than 20,000 is 415. Hence, in our notation above, $x = 415$.

Therefore, the estimate of the proportion below the poverty line is

$$E(\theta|x) = \frac{416}{1002} = 0.4152$$

$$Var(\theta|x) = 0.000242$$

The standard deviation of the posterior distribution of θ is the square root of the variance, which is 0.0156.

Figure 3.2 shows the posterior distribution of θ. We can see that it is very closely distributed around the mean. From this probability density function (pdf), we can calculate any interesting probability about θ. A typical one is a 95% **credible interval**. This would be the two values, θ_L and θ_H, such that

$$P(\theta_L < \theta < \theta_H) = 0.95$$

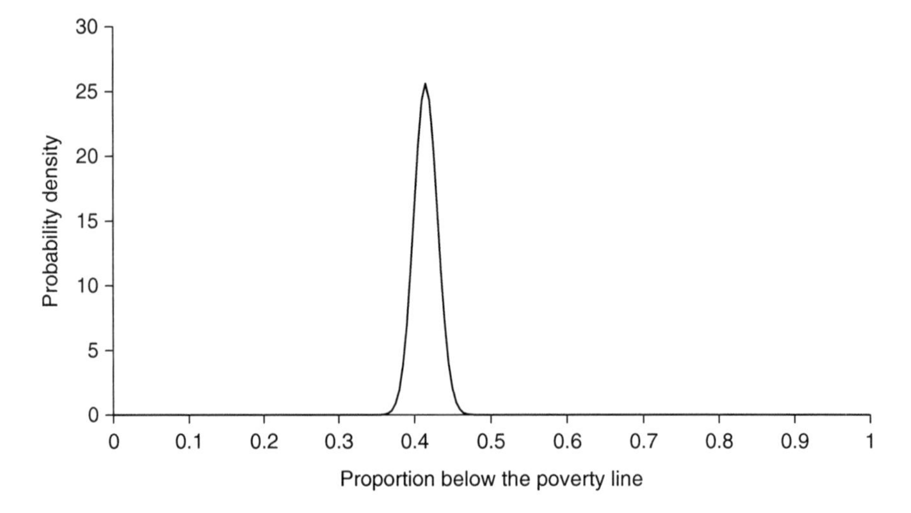

Figure 3.2 Posterior distribution of θ, the proportion below the poverty line

In this case, $\theta_L = 0.39$ and $\theta_H = 0.45$ so that there is 95% probability that the 'true' value of θ is in the range 0.39 to 0.45. θ remains unobservable, but once we have found a posterior distribution based on some data, we can begin to make informed probability statements about it.

Let's be clear about the meaning of all this. If before we had the data we had no knowledge at all about θ (which is why we used the uniform distribution), then after we obtain the data, Bayes' theorem gives us the posterior distribution, from which we can make probability statements and find estimates such as the mean and standard deviation of the distribution. Note that a

95% credible interval before we had the data was 0.025 to 0.975 (95% of the uniform distribution is within these limits). Obtaining these data has considerably narrowed the 95% credible interval range.

There are many possible ways of choosing a 95% credible interval. There is not 'the' 95% credible interval, only 'a' 95% interval. For example, for the uniform distribution, the range 0 to 0.95 also contains 95%, as does the range 0.05 to 1.

The two most commonly used versions for a credible interval are called 'equal tails' and 'highest density range (HDR)'. The equal tails one (which is what we have used) finds θ_L and θ_H such that

$$P(\theta < \theta_L) = 0.025$$

$$P(\theta > \theta_H) = 0.025$$

so that exactly 2.5% of the distribution is to the left and 2.5% to the right of the interval bounds, with 95% in between. Instead, the HDR interpretation finds the smallest interval such that 95% of the distribution is within that interval. This has the particular interpretation that every value in this range has a greater probability than points outside this range. We will consider how to compute the HDR in practice later.

Finally, we note that there is nothing sacrosanct about 95% – it could be any probability, but typical choices for summarizing the distribution are 90%, 95%, and 99%.

Equation (3.4) may be considered as an *estimator* for θ. An estimator is some function of the data, so that when the actual data values are substituted in, we get a specific *estimate*. In this case, the specific estimate is 0.4152.

Let's look at the properties of this estimator. Consider again what θ represents – it is the (unknown) proportion of people in the working population who receive an income below the poverty line. Another way to think of this is that if you selected a person at random from the working population, the probability that they would be below the poverty line is θ. Now look at the estimator and consider what happens as $n \to \infty$. For large n, there is negligible difference between $\frac{x}{n}$ and $\frac{x+1}{n+2}$. So the estimator 'at infinity' (i.e. large numbers) is equivalent to the probability in which we are interested (i.e. θ). This can also be seen by the fact that the variance tends to 0 as n tends to infinity. Hence, the estimator gets closer and closer to the 'true value' with greater probability as the sample size increases. Such an estimator that converges probabilistically to the 'true value' of the parameter as the sample size increases is called a *consistent estimator*.

Let's contrast consistency with another possible property of estimators. We know from the binomial distribution that $E(x|\theta) = n\theta$. Hence,

$$E\left(\frac{x+1}{n+2}\,\middle|\,\theta\right) = \frac{n\theta+1}{n+2} = \theta\left(\frac{n}{n+2}\right) + \frac{1}{n+2}$$

In other words, the expected value of the estimator is not equal to what it is supposed to be estimating (i.e. θ, though it is close and for large n the difference becomes negligible). Such an estimator is said to be *biased*. An *unbiased estimator* (t) would be where $E(t|\theta) = \theta$. In fact, $t = x/n$ is an unbiased estimator.

Consider for a moment what 'unbiased' really means in a physical sense. We take a sample of n individuals and compute the estimate from the estimator, say t, and we record the answer, say t_1. Then, we take another sample of n individuals and do the same, and we get t_2. We keep doing this again and again, a large number of N times, so that finally we have $t_1, t_2, \ldots, t_N$. Now, we take the mean (average) of all these and get $\bar{t}$. If t is an unbiased estimator and N is large enough, then $\bar{t}$ will have a high probability of being very close to the value of the unknown parameter (in the limit if N is infinity, then the probability is 1). So an 'unbiased' estimator is one that 'on the average' is equal to the true value of the parameter.

In Bayesian statistics, we do not worry much about being unbiased. The fact is that we only ever have *one sample*, so what happens with all these other hypothetical samples that *might have been drawn* is not of any particular interest. If we really did have two samples, obtained in the same manner, then we would join them together to make one bigger sample. Consistency is the far more important property. We want to know that if we do have large sample sizes, then there is a high probability that the parameter estimate will be right.

The Borgonian Minister for Social Justice became aware of these findings. He strongly objected. He said that he is really convinced that the true proportion of people below the poverty line is about 5%, and that the finding of around 40% is 'Fake news, and everyone knows it.' How can we take account of the minister's concerns? We have a way to do this through the prior distribution. We assumed that nothing was known about the proportion below the poverty line, so we gave it a uniform prior. But if the minister is so convinced that it is around 5%, then we should choose a prior that reflects this. We know from the Beta distribution $B(a,b)$ that its mean is $\frac{a}{a+b}$. Let's choose a and b to reflect the minister's beliefs. For example, $a = 5$ and $b = 95$ would give a mean of 0.05 and a small variance for the prior distribution (calculate it).

Now we will have the prior distribution:

$$p(\theta) \propto \theta^4(1-\theta)^{94}$$

Hence, the posterior distribution is

$$f(\theta|x) \propto \theta^{x+4}(1-\theta)^{n-x+94}$$

This is a $Beta(x + 5, n - x + 95)$ distribution, with the observed $x = 415$.

Hence, the mean estimate is

$$\frac{x+5}{n+100} = 0.3818$$

The 95% credible interval for this distribution (equal tails) is 0.35 to 0.41. Hence, in spite of the minister's very firm belief (the prior 95% credible interval is 0.02 to 0.10), the data have spoken and have overridden the opinion of the minister.

Although there are very different priors, the posterior distributions are still not particularly different. However, it is important to note that this happens because the sample size ($n = 1000$) is large. If we had a sample size of, say, 50 or 100, the story may be different. (This also helps us understand the importance of the consistency property of estimators.)

It is left to the reader to repeat all of the above, but with a sample size $n = 100$.

Conjugate Priors

Consider again the example where the likelihood is binomial and the prior is Beta:

Binomial likelihood: $f(x|\theta) = \binom{n}{x}\theta^x(1-\theta)^{n-x}$

Beta prior: $p(\theta) \propto \theta^{a-1}(1-\theta)^{b-1}$ for known values a and b

You may notice that although the two distributions, binomial and Beta, are quite different, their functional forms with respect to θ are the same. So, when we find the posterior,

posterior $\propto$ *likelihood* $\times$ *prior*

$p(\theta|x) \propto \theta^{x+a-1}(1-\theta)^{n-x+b-1}$

Here the prior is a Beta distribution Beta(a, b) and the posterior is also Beta, Beta($x + a$, $n - x + b$). The posterior has the same distribution as the prior, but the new parameters now depend on the observed data. This is an example when the prior is said to be *conjugate* to the likelihood. The data simply update the parameters of the prior.

Let's consider another example, where the likelihood is a Poisson distribution with mean μ:

Poisson likelihood: $f(x|\mu) = \frac{\mu^x}{x!}e^{-\mu}$

Gamma prior: $p(\mu) = \frac{1}{\Gamma(a)}\mu^{a-1}e^{-\mu}$ for known $a > 0$

Here the prior is Gamma, with parameters shape a and rate 1. Hence, the posterior is

$p(\mu|x) \propto \mu^{x+a-1}e^{-2\mu}$

Comparing with Chapter 2 (Equation 2.36), this is also the Gamma distribution with shape $x + a$ and rate 2. Here, the Gamma distribution is conjugate to the Poisson, so that when the two pdfs multiply, the resulting posterior is also Gamma.

Conjugates are mentioned here not because they are necessary, but because they are convenient. They give rise to a situation when the prior and the posterior are distributions of the same type and only the parameters are updated with the parameters of the posterior dependent on the observed data. This is mathematically convenient and helps in interpretation. However, in reality, as we shall see, we compute posterior distributions numerically, with the aid of computer software, so there is no intrinsic reason why conjugate priors need to be chosen. Priors should be chosen to best represent prior knowledge. In many circumstances, there will not be a conjugate prior.[2]

[2]There is a nice graphical summary of conjugate priors on www.johndcook.com/blog/conjugate_prior_diagram/

Inference About the Mean and Standard Deviation of the Normal Distribution

Now apart from making an inference about one aspect of the distribution of income, we consider how to summarize the distribution as a whole. One obvious summary is to find the mean of all the data (the sum divided by 1000). However, if we are interested in income *distribution*, we also need to know how much departure there is from this mean. For that we could use the standard deviation (the square root of the variance). However, the values we obtain will be solely with respect to the 1000. We need to have some idea about the population as a whole.

To get any further, we have to make a number of abstractions. We will assume that the random variable 'income' is continuous. In fact, we know that the salary of a Borgonian is not really on a continuous scale. The Borgonian currency is in b€ and bC (Borgonian cents). Although theoretically, and in the world of finance, one could come across numbers like 75,789.159393 b€ on a computer, in everyday life, items and salaries are in just Borgonian euros and cents, and you will never find something costing, for example, $\sqrt{2}$ or π b€. In reality, the random variable 'income' is discrete. However, we will abstract away from this and take it as being continuous.

A second assumption is that the distribution of income is such that the mean is itself meaningful (no pun intended) as a 'model' of the distribution. For example, suppose that the distribution were bimodal (it has two peaks) or is U-shaped (see Figure 2.6), then the mean would not be a very useful summary. Consider this set of numbers: 1, 1, 1, 1, 1, 1, 9, 9, 9, 9, 9, 9. The mean is 5. Yet none of the numbers is equal to (or even 'close to') 5. It is a pretty useless summary of this set of numbers.

A literature search will quickly reveal that there has been a lot of research on income distributions. A good model for the distribution of income turns out to be the 'lognormal' distribution (Chapter 2). This is where y is a random variable such that $\log(y) \sim normal(\mu, \sigma)$. In other words, the logarithm of y has a normal distribution with mean μ and standard deviation σ, so that y itself has the lognormal distribution.

Before we look at this, we will adopt a simpler model of income distribution which is to say that the distribution of income is itself *normally* distributed (rather than worry now about logs). There is a good reason why the lognormal is used in practice for income distribution, but we will not be concerned with that right now. So we will assume income, which we will denote as y, is a random variable with a normal distribution $normal(\mu, \sigma)$. Here, the y are observable, and the μ and σ are unobservable.

Instead of writing 1000, we will use n as before. So we will have observations $y_1, y_2, \dots y_n$ on n independent random variables, each with the $normal(\mu, \sigma)$ distribution. The observations can be considered independent because of the random sampling. Hence their conditional joint distribution may be written as

$$f(y_1, y_2, \dots y_n|\mu, \sigma) = f(y_1|\mu, \sigma)\, f(y_2|\mu, \sigma) \dots f(y_n|\mu, \sigma)$$

where

$$f(y_i|\mu,\sigma) = \frac{1}{\sigma\sqrt{2\pi}} \exp\left(-\frac{1}{2}\left(\frac{y_i - \mu}{\sigma}\right)^2\right), \; i=1, 2, \dots, n$$

This is the 'likelihood'. It is a model of the 'data' assuming that we knew the values of unobservables, μ and σ. Taking one step further, the joint distribution can be rewritten as

$$f(y_1, y_2, \ldots, y_n | \mu, \sigma) = \left(\frac{1}{\sigma\sqrt{2\pi}}\right)^n \exp\left(-\frac{1}{2\sigma^2}\sum_{i=1}^{n}(y_i - \mu)^2\right) \quad (3.5)$$

The quantities of interest are μ and σ, and our goal is to be able to make statements about these quantities once we have the actual data (the observed values of $y_1, y_2, \ldots, y_n$). How can we do this?

The distribution of income from the sample of 1000 is shown in Figure 3.1. Unfortunately, it does not seem to satisfy our assumption that the distribution is normal; however, we will ignore that problem and come back to it later. The likelihood is given in Equation (3.5). However, we require the posterior distribution:

$$f(\mu, \sigma | y_1, \ldots, y_n) \propto f(y_1, \ldots, y_n | \mu, \sigma)\, p(\mu, \sigma)$$

Remember, this is simply

posterior distribution $\propto$ *likelihood* $\times$ *prior*

(We repeat this a lot because it is fundamental!)

How shall we choose the prior distribution $p(\mu, \sigma)$? Let's make the simplifying assumption that we 'know nothing'. In other words, μ and σ are any numbers at all, although it is essential that $\sigma > 0$. The uniform distribution would have been suitable as a prior distribution – except that it has a finite range. Since the unknown parameters have infinite ranges, the uniform distribution is not possible. However, here we can do a trick. We can 'pretend' that the prior distributions are of the form

$$f(\mu) \propto 1$$

$$-\infty < \mu < \infty$$

$$f(\sigma) \propto \frac{1}{\sigma}$$

$$\sigma > 0$$

Clearly, these prior distributions are not real probability distributions, since they certainly do not integrate to 1 (obviously the integrals are infinite). We can nevertheless try using them, and 'see what happens'. Such priors are called *improper*. Improper priors sometimes give valid posteriors. Note that the distribution for σ is equivalent to $\log\sigma$ being uniformly distributed over the range $-\infty < \log\sigma < \infty$. (This follows from distribution theory, but is beyond the scope of this book.)

We shall use these improper prior distributions and see what we get. Hence, the joint posterior distribution of μ and σ is

$$f(\mu, \sigma | y_1, y_2, \ldots y_n) \propto \left(\frac{1}{\sigma}\right)^{n+1} \exp\left\{-\frac{1}{2\sigma^2}\sum_{i=1}^{n}(y_i - \mu)^2\right\}$$

(Readers uninterested in the mathematics can skip to the next section.)

For notational convenience, it is a convention to write the observations $(y_1, y_2, \dots y_n)$ as $\mathbf{y}$. To obtain the posterior distribution of μ, we can integrate over σ:

$$f(\mu \mid \mathbf{y}) \propto \int_0^\infty \left(\frac{1}{\sigma}\right)^{n+1} \exp\left\{-\frac{1}{2\sigma^2}\sum_{i=1}^{n}(y_i-\mu)^2\right\} d\sigma$$

The distribution for μ obtained by integrating out σ is sometimes called the 'marginal posterior' distribution for μ. There is no new concept here, this term is only mentioned in case the reader comes across it in other literature. As we saw in Chapter 2, if there is a joint distribution of several variables, then the distribution of any subset of these can be obtained by integrating out the ones that are not of immediate interest. This is exactly the same here; at the moment we are interested in the distribution of μ irrespective of σ.

Continuing, with some manipulation it is possible to show that

$$\sum_{i=1}^{n}(y_i-\mu)^2 = \sum_{i=1}^{n}(y_i-\bar{y})^2 + n(\mu-\bar{y})^2$$

where $\bar{y} = 1/n\sum_{i=1}^{n} y_i$ is the mean of the observations. Finally, we use the notation

$$s^2 = \frac{1}{n-1}\sum_{i=1}^{n}(y_i-\bar{y})^2$$

which is the (unbiased) sample variance.

Using all this, we can rewrite the posterior distribution as

$$f(\mu,\sigma \mid \mathbf{y}) \propto \left(\frac{1}{\sigma}\right)^{n+1} \exp\left\{-\frac{1}{2\sigma^2}(n-1)s^2\right\} \times \exp\left\{-\frac{1}{2}\left(\frac{\mu-\bar{y}}{\sigma/\sqrt{n}}\right)^2\right\}$$

Consider the rightmost expression (after the × symbol) – this is the pdf for the *normal* $\left(\mu, \frac{\sigma}{\sqrt{n}}\right)$ distribution (apart from the normalizing constant). Since we know that

$$f(\mu, \sigma \mid \mathbf{y}) \propto f(\mu \mid \sigma, \mathbf{y})\, f(\sigma \mid \mathbf{y})$$

it must be the case that

$$f(\mu \mid \sigma, \mathbf{y}) \propto \exp\left\{-\frac{1}{2}\left(\frac{\mu-\bar{y}}{\sigma/\sqrt{n}}\right)^2\right\}$$

$$f(\sigma \mid \mathbf{y}) \propto \left(\frac{1}{\sigma}\right)^{n+1} \exp\left\{-\frac{1}{2\sigma^2}(n-1)s^2\right\}$$

The first of these is therefore the posterior distribution of μ conditional on σ. It is interesting that the conditional distribution of μ is a normal distribution, but this is not much help unless we happen to know σ, which of course we don't.

The second gives us the posterior distribution of σ. Note that this is only dependent on the data; in fact, provided we know the sample variance, we don't even need the original data. As we might intuitively expect, the posterior distribution of σ depends on the sample standard deviation s. Here the sample variance s^2 is said to be a *sufficient statistic* for σ. What this means is that we can throw away all the data and simply retain the value of s^2, and this would be sufficient to know the posterior distribution of σ.

By re-parameterizing in terms of the variance by writing $\tau = \sigma^2$, we can obtain the distribution of τ as

$$f(\tau|\mathbf{y}) \propto \left(\frac{1}{\tau}\right)^{\frac{n}{2}+1} \exp\left\{-\frac{\frac{1}{2}(n-1)s^2}{\tau}\right\}, \quad \tau > 0$$

Comparing with Chapter 2, this is the inverse Gamma distribution with $\alpha = \frac{n}{2}$ and $\beta = \frac{1}{2}(n-1)s^2$. (Note that the re-parameterization in terms of the variance is not simple substitution but involves also multiplying by the derivative $\frac{d\sigma}{d\tau} = \left(\frac{1}{\tau}\right)^{1/2}$).

To find the posterior distribution of μ alone, we need to integrate out σ from the joint distribution:

$$f(\mu|\mathbf{y}) \propto \int_0^\infty \left(\frac{1}{\sigma}\right)^{n+1} \exp\left\{-\frac{1}{2\sigma^2}\sum_{i=1}^{n}(y_i-\mu)^2\right\} d\sigma$$

Again, we can make the substitution $\tau = \sigma^2$, and make use of the fact that we know from the inverse Gamma distribution that

$$\int_0^\infty \left(\frac{1}{u}\right)^{\alpha+1} \exp\left\{-\frac{\beta}{u}\right\} du = \frac{\Gamma(\alpha)}{\beta^\alpha}$$

Then,

$$\begin{aligned} f(\mu|y) &\propto \left[\frac{1}{\sum_{i=1}^{n}(y_i-\mu)^2}\right]^{\frac{n}{2}} \\ &= \left[\frac{1}{(n-1)s^2 + n(\mu-\bar{y})^2}\right]^{\frac{n}{2}} \\ &= \left(1+\frac{t^2}{n-1}\right)^{-\frac{n}{2}} \end{aligned}$$

where

$$t = \frac{\mu-\bar{y}}{s/\sqrt{n}}$$

From Chapter 2, we can see that this is a Student t distribution with $n-1$ degrees of freedom.

Inference About the Mean and Standard Deviation of the Normal Distribution With Improper Priors

We, therefore, have arrived at the following result. If we *know nothing* about the mean and variance of the normal distribution and have n independent observations drawn on the distribution, then the posterior distribution of the variance has an inverse Gamma distribution, and the posterior distribution of the mean is a Student t distribution. The second statement is the Bayesian equivalent of the well-known 'Student's t test' of classical statistics. We can see that this test, from a Bayesian perspective, is based on the (dubious) assumption that we 'know nothing' about the unknown parameter values – for example, the prior knowledge would include the notions that the mean income of 100,000,000 b€ is just as likely as 10,000 or –10,000 in the case of the mean, and similarly for the log of the variance. The assumption is that any value from minus to plus infinity is equally likely.

The Student t distribution is almost equal to the normal distribution for degrees of freedom of around 30 or more. Since $n = 1000$, we can use the normal distribution instead, with mean and standard deviation $s/\sqrt{n}$. (Recall also the central limit theorem introduced in Chapter 2.)

From these data, mean income $\bar{y} = 25{,}552$, and $s = 14{,}136$. Using these, we can plot the posterior distribution of μ, the *mean income*, as shown in Figure 3.3. Recall that the prior distribution is an improper distribution that is flat over the whole real line. From the normal distribution (or equivalently the t distribution with high degrees of freedom), we know that

$$P(-1.96 < t < 1.96) = 0.95$$

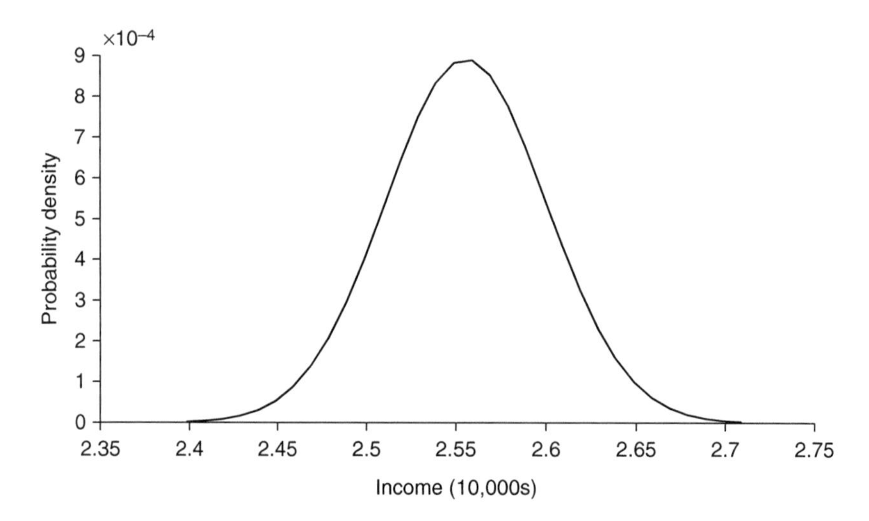

Figure 3.3 Posterior distribution of mean income

Hence,

$$P\left(-1.96 < \frac{\mu - \bar{y}}{s/\sqrt{n}} < 1.96\right) = 0.95$$

and therefore

$$P\left(\bar{y}-1.96\frac{s}{\sqrt{n}}<\mu<\bar{y}+1.96\frac{s}{\sqrt{n}}\right)=0.95$$

This is a 95% credible interval for the value of μ, simply the posterior probability that μ is between these two limits which are 24,676 to 26,428. Similarly, we could find a 99% credible interval (or any other that we wished – left as an exercise for the reader).

To examine the posterior distribution of the variance, let's work in units of income of 10,000 rather than single units. Then, $s = 1.4136$, $\alpha = n/2 = 500$, and $\beta = 0.5(n-1)s^2 = 998.1537$. These numbers are still very large to be able to efficiently compute the pdf, so instead, we can compute the log of the pdf, which is

$$\log f(\tau|y) \propto -\left(\frac{n}{2}+1\right)\log(\tau)-\frac{(n-1)s^2}{2\tau}$$

where τ is the variance. This is plotted in Figure 3.4.

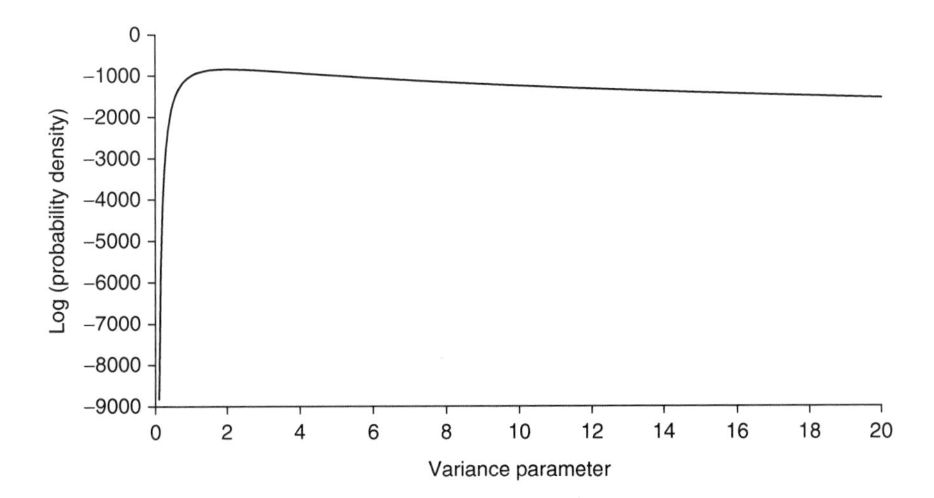

Figure 3.4 Log probability density for the inverse Gamma distribution of the variance τ

This distribution sharply rises to its maximum value of 2, and then slowly tails off for values greater than 2. Recall that the observed standard deviation $s = 1.4136$ (now in units of 10,000), and, therefore, the sample variance is $s^2 = 1.9973$. Hence, the mode of the posterior distribution is very close to that of the observed sample. Also, the expected value (mean) of the inverse Gamma distribution is (Chapter 2, Equation 2.47):

$$\frac{\beta}{\alpha-1}=2.003$$

which is also very close to the sample variance. The variance of the inverse Gamma distribution is

$$\frac{\beta^2}{(\alpha-1)^2(\alpha-2)}=0.008$$

and hence the standard deviation of the distribution is 0.0896.

Let's summarize where we are. We started with 1000 observations on income. That's a lot of information! Instead, we now have a model of income distribution which can be summarized in a few lines. So remembering that the notation y refers to income,

$$y \sim normal(\mu, \sigma)$$

$$t = \frac{\mu - \bar{y}}{s/\sqrt{n}} \sim Student\ t(n-1)$$

$$\tau = \sigma^2 \sim inverse\_gamma\left(\alpha = \frac{n}{2}, \beta = \frac{1}{2}(n-1)s^2\right)$$

Assuming that this model were a good characterization of the distribution of income, we would be able to use this model to make statements or predictions about income. However, how do we know that the model is a good one? One thing we can do is to generate random observations from the model and then see how well the resulting predicted distribution of income, the one predicted by the model, fits the actual distribution from our original 1000 observations. Statistical software would normally allow this possibility, and we will show how it can be done later. Basically, pseudo-random samples would be drawn from the joint distribution of (μ, σ) and then these used to draw pseudo-random samples from y. Figure 3.5 shows the histogram of 8000 pseudo-random observations on the above model.

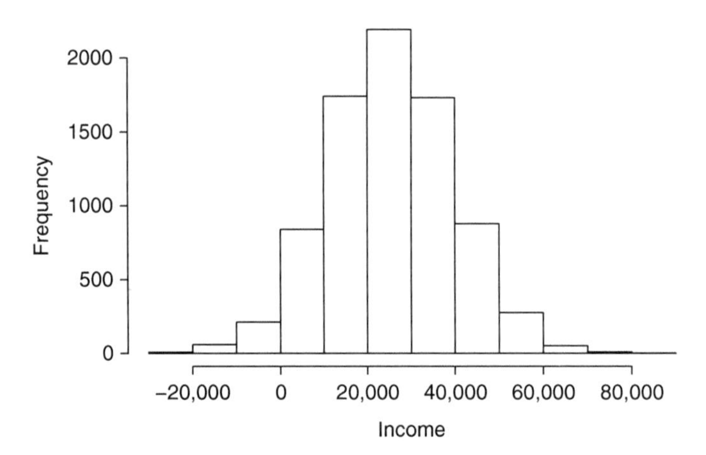

Figure 3.5 Histogram of predicted income of 8000 observations from the model for income distribution

Of course, as the model stipulates, this looks very much like a sample from a normal distribution. It hardly resembles the income distribution from the observed data shown in Figure 3.1. Moreover, from this predicted distribution, we can also consider the predicted proportion below the poverty line (20,000 b€). Here, the probability of having an income of less than 20,000 b€ comes out to be 0.36. This is outside the original 95% credible interval we had computed (based on a uniform prior distribution) and just inside the 95% credible interval based on the prior that took into account the views of the Borgonian minister.

There are at least two aspects of our model that could be improved. As we remarked earlier, the likelihood model that we chose (a normal distribution for the distribution of income) is not appropriate. Income distributions are usually skewed, with a long tail to the right expressing the fact that there are some individuals who have incomes that are extreme compared to the mean, mode, or median, as can be seen in Figure 3.1. The second problem is that our prior distributions for the mean and variance did not take into account basic facts – like incomes cannot be negative (well actually in practice they unfortunately can be – but not in the sense meant here!), whereas our prior distribution for μ gave equal weight to negative as to positive values. (Indeed the predicted distribution for income includes some negative values.) Moreover, the prior distribution for the standard deviation was an improper one, making all possible non-negative values equally likely.

Inference on the Lognormal Distribution Using R and Stan

Let's fix the first problem – we will make the distribution of income (the likelihood) be a log-normal distribution. From Chapter 2, this has the following pdf:

$$f(y|\mu,\sigma) = \frac{1}{y\sigma\sqrt{2\pi}} \exp\left(-\frac{1}{2}\left(\frac{\log(y)-\mu}{\sigma}\right)^2\right)$$

(Note that this is almost the same as the normal distribution substituting y by $\log(y)$ except from distribution theory we need the $1/y$, which is the derivative of $\log(y)$.)

Let's leave the prior distributions for the non-observables μ and σ to be improper, for the moment. Recall also from Chapter 2 that μ and σ are not the mean and variance of y, they are the mean and variance of the distribution of $\log(y)$. For convenience, let's write here again that

$$E(y|\mu,\sigma) = \exp\left(\mu + \frac{\sigma^2}{2}\right)$$

$$Var(y|\mu, \sigma) = (\exp(\sigma^2) - 1)(\exp(2\mu + \sigma^2))$$

Now we could try to do all the mathematics again and attempt to derive the posterior distributions for μ and σ. Let's instead do this an easier way and let software come to the rescue. First write down what we know:

Box 3.1

The data:
$n = 1000$
y is a vector of length n

The parameters
μ and $\sigma > 0$

Transformed parameters
$mn = \exp\left(\mu + \frac{\sigma^2}{2}\right)$ is the mean

$sn = \sqrt{\left(\exp\left(\sigma^2\right) - 1\right)\left(\exp(2\mu + \sigma^2\right)}$ is the standard deviation

The model:
prior distributions of μ and $\sigma > 0$ are improper
income ~ *lognormal*(μ, σ)

Let's write exactly the same information in a different way:

Box 3.2

```
data {
  int<lower=0> n;
  vector[n] y;
}

parameters {
  real mu;                          //parameters of lognormal
  real<lower=0> sigma;
}

transformed parameters{
  real mn;                          //mean of lognormal
  real v;
  real sn;                          //standard deviation of lognormal

  v = sigma*sigma;

  mn = exp(mu + v/2);
  sn = sqrt((exp(v)-1)*exp(2*mu +v));
```

```
}

model {
  y ~ lognormal(mu,sigma);
}
```

Compare Box 3.1 with Box 3.2. The data block simply specifies the data – note that it is enforcing what we know must be true: that the number of observations, n, is non-negative '<lower=0>', and that y is a vector of real values of that length, 'vector[n] y'. Next, it specifies the parameters, noting that σ cannot be negative. Here 'real' means that the values of μ and σ are continuous, with the only restriction that $\sigma > 0$. Recall that our interest is not really in the parameters of the lognormal distribution μ and σ but in the mean and standard deviation of the distribution as given above. The 'transformed parameters' is used to compute these quantities. Finally, the 'model' block specifies the likelihood, the distribution of income conditional on the unknown parameters. Note that we have not specified the prior distributions of μ and σ since by default these are taken as improper priors (with the stipulation that $\sigma > 0$ enforced) unless otherwise specified – which we will come to below.

This is a computer program! It can be used, for example, to compute the posterior distributions for the unobservables. The program is specified in a language called Stan (https://mc-stan.org; Stan Development Team, 2011–2019). This type of software is what has made Bayesian statistics possible in recent decades – since if for every problem we had to mathematically derive the posterior distributions, Bayesian statistics would be impractical, especially if there were no conjugate priors. First, it is very tedious to have to do this, and, second, it is rarely even possible to be able to carry out all the integrations and find closed formulae (as we did above). Instead, numerical methods are used to solve the integrals and, effectively provide samples from the joint distributions of all the parameters, besides other things. We will come back to how we actually use and invoke this program shortly. But now let's use it and see the results we get:

```
Inference for Stan model: model1income.
4 chains, each with iter=4000; warmup=2000; thin=1;
post-warmup draws per chain=2000, total post-warmup draws=8000.

          mean se_mean     sd     2.75%     97.5% n_eff Rhat
mu       10.01    0.00   0.02      9.98     10.05  7208    1
sigma     0.52    0.00   0.01      0.50      0.55  5955    1
mn    25593.62    5.24 444.69 24752.35 26483.90  7201    1
sn    14353.43    6.32 501.24 13442.65 15373.95  6293    1

Samples were drawn using NUTS(diag_e) at Sat Feb 16 12:09:50 2019.
For each parameter, n_eff is a crude measure of effective sample
size, and Rhat is the potential scale reduction factor on split
chains (at convergence, Rhat=1).
```

The last two columns give information about the convergence of the simulation carried out in order to obtain the results. Except for noting that the 'Rhat' values should be all equal to 1 in order to show useful convergence, this is a topic we will come back to in Chapter 7.

A simulated posterior distribution for each parameter is obtained, and the results show their means, standard errors, standard deviations, and the equal tail 95% credible intervals. For example, the posterior distribution of the mean income 'mn' has an expected value 25,596 b€ with standard deviation 445 b€, and a 95% credible interval between 24,752 and 26,483 b€. The distribution of the standard deviation of income 'sn' has mean 14,353, and a 95% credible interval between 13,443 and 15,373.

The Stan web page shows that there are many interfaces to Stan through languages such as R, MATLAB, Python, and so on. In this book, we will show only the R interface. Note that this only relates to how the Stan program is set up, accessed, and executed but not the Stan program itself. The Stan program is the same whether called from R, MATLAB, or whatever.

To call the above program using R, the following was used:

Box 3.3

```
mydata <- list(n = 1000, y = income)
```

This simply establishes a list called 'mydata' which contains two elements, 'n' the number of observations and the specification of 'y' as equivalent to an already existing variable called 'income' which contained the 1000 observations. Then, the program above is in a file called, for example, 'model1income.stan', and the following R interface function is invoked:

Box 3.4

```
fit  <- stan (file = "model1income.stan", # Stan program
                  data = mydata, # named list of data
                  chains = 4, # number of Markov chains
                  iter = 4000, # total number of iterations per chain
                  cores = 4, # number of cores)
```

The variable 'fit' is the output (it could be any variable name), and the name of the function is 'stan'. The keyword 'file' references the file which contains the Stan program. (In the online program, we use a slightly different method to specify the program.) The keyword 'data' specifies the list or 'data frame' where the data are stored. So from this, the Stan program gets the value of 'n' and also the vector 'y'. The keyword 'chains' specifies how many independent simulations are going to be executed, and each one will have 4000 iterations (the keyword 'iter'). The keyword 'cores' specifies how many cores on the computer that you are using should be used if possible. The computer used in this case had four cores. Ideally, each independent 'chain' should be executed in parallel across the four cores.

In order to print the output from the call to the 'stan' function, we can use

Box 3.5

```
print(fit, pars=c("mu", "sigma","mn","sn"), probs=c(.0275,.975))
```

Here, we print the output from the call ('fit'). The 'pars' keyword specifies the list of parameters that we want to list, and the 'probs' keyword specifies the list of probability levels that we want output for each of the parameters (in this case, we wanted the equal tail 95% credible intervals). Of course, these probabilities can be whatever is desired – for example, including 0.5 in the list would give the medians of the posterior distributions.

We can extract further information from 'fit' by calling

Box 3.6

```
e <- extract(fit)
```

The variable 'e' (which could have been any variable name) is now a structure that contains the actual simulated distributions of the parameters specified by the keyword 'pars'. In this case, 'e$mn' contains all the values of the simulated distribution of the 'mn' parameter, and 'e$sn' contains the 'sn' parameter distribution, and using the R 'hist' and 'plot' functions, we can plot these distributions).

Box 3.7

```
hist(e$mn, main = "", xlab = "mean income", freq = FALSE,
         ylab = "probability density", axes = FALSE)
par(new = TRUE)
plot(density(e$mn), main="",xlab="",ylab="")
```

The first line plots the histogram of 'mn'. The 'xlab' and 'ylab' keywords are specifications of axis labels. The 'freq' keyword specifies that density rather than frequency should be plotted on the y-axis, and 'axes' being FALSE will prevent the axes of the histogram and associated labelling. This is because we want the axes plotted in the 'plot' command. The density function will compute a smooth pdf from the data, and 'plot' will plot it. The 'par(new=TRUE)' is just a way to overlay one graph on top of the other. The result is shown in Figure 3.6.

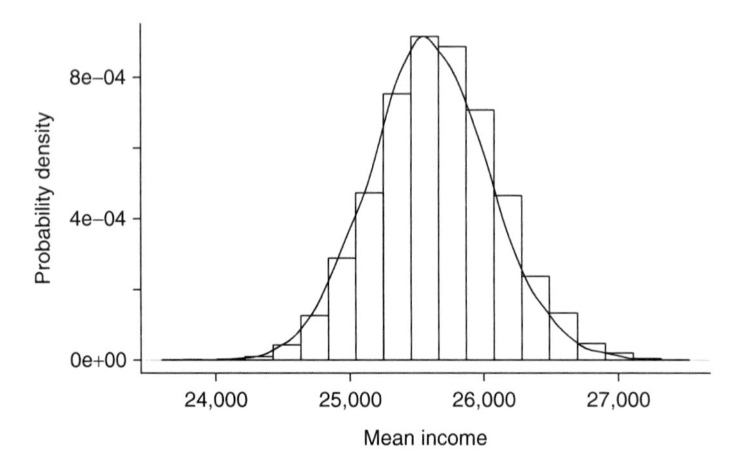

Figure 3.6 Probability density for mean income distribution based on a sample of 8000 simulated observations from the posterior distribution

Let's add one more block to the Stan program and execute the program again:

Box 3.8

```
generated quantities{
  real income_new;

  income_new = lognormal_rng(mu,sigma);

}
```

The 'generated quantities' block will allow simulation of new variables from the Bayesian solution. Recall that when we used the original model for income, the predicted distribution of the model was normal and looked quite unlike the actual observed distribution of income. Now, given our new model, let's examine how the predicted distribution of income looks. In this block, the values of 'mu' and 'sigma' are sampled from their joint posterior distribution and then used to generate a random observation on the lognormal distribution. 'lognormal_rng' does this random generation. Hence, 'income_new' will end up as a vector of 8000 values based on 8000 pseudo-random simulations on 'mu' and 'sigma'. Now, if we run the model again, and extract using

Box 3.9

```
e <- extract(fit)
```

we will obtain this predicted distribution as 'e$income_new':

Box 3.10

```
hist(income, main = "", xlab = "",
   freq = FALSE, ylab = "", axes = FALSE)
par(new = TRUE)
plot(density(e$income_new), main="",
   xlab="mean income",ylab="probability density")
```

This plots the original histogram of income and the density function of the posterior predicted income from the model. The result is shown in Figure 3.7. The theoretical distribution matches the mode and skewness of the original observations, although it is not so close at the higher ranges of income. Nevertheless, this is a distribution of the same form as the original, rather than a completely different distribution, so there is some progress.

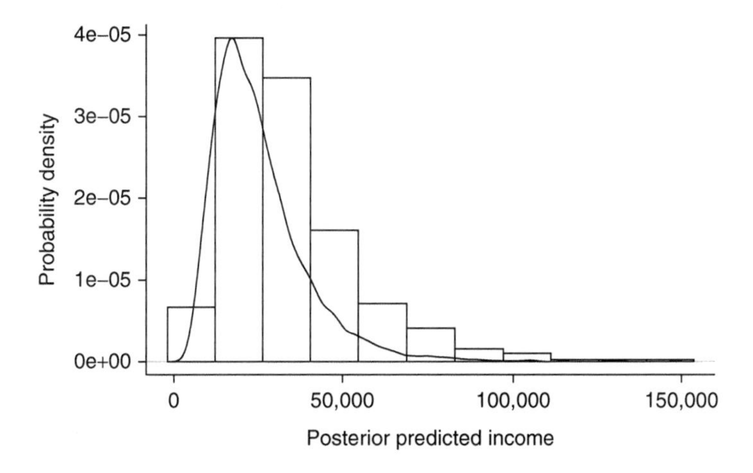

Figure 3.7 The original histogram of income is shown superimposed on the posterior predicted distribution of income (*solid line*)

Using the posterior distribution, we can also calculate any probabilities that we like. In particular, let's now see the probability of an income being below the poverty line. This is

$$P(yincome < 20{,}000|\mu, \sigma)$$

In terms of the R interface to Stan, this would be

Box 3.11

```
mean(e$income_new < 20000)
```

(It is the sum of all the times that the income is less than 20,000 divided by the number of times.) This computes the proportion of values of 'e$income_new' that are less than 20,000 (the cut-off income for the poverty line). The answer comes out to be 0.42. Recall that the actual value in the observed income is 0.415, which was also the estimate we obtained from the original analysis using the binomial distribution. This is much better than the 0.36 found from the previous model.

Prior Distributions for the Parameters

Up to now, we have used improper prior distributions for the parameters. Let's consider this, though. We think we know nothing about the mean income, but in fact we do know some obvious things. If the poverty line is 20,000 b€, it is extremely unlikely that the *median* income is going to be less than that. On the other hand, from what we know about Borgonian society, prior to collecting any data, it would be highly unlikely that the median income is going to be greater than 120,000 b€. (Remember, we are talking here about *median* income.) We could say that we are 99% sure that the median income is going to be between these two limits. Hence, we could write, as a first step in choosing appropriate prior distributions for μ for the lognormal distribution, recalling that we are working in units of 10,000,

$$P(2 < median < 12) = 0.99$$

However, we know from the properties of the lognormal distribution that $median = \exp(\mu)$ (Chapter 2).

Hence,

$$P(2 < \exp(\mu) < 12) = 0.99$$

Or

$$P(\log(2) < \mu < \log(12)) = 0.99$$

Let's therefore give μ a prior normal distribution that satisfies this property. Since for the standard normal distribution, 99% lies between the limits ±2.58, some simple arithmetic shows that we require a normal distribution with mean 1.589 and standard deviation 0.347. This is so because, if we let $\mu \sim normal(\mu_0, \sigma_0)$, then we require

$$\mu_0 - 2.58\sigma_0 = \log(2)$$

$$\mu_0 + 2.58\sigma_0 = \log(12)$$

The σ parameter is harder to argue in this way. However, rather than use an improper prior for σ, we can use an actual distribution but with a very wide set of possible values. Recall from Chapter 2 (Equation 2.45) that the Cauchy distribution has infinite variance, so that very high values of a Cauchy random variable can have probabilities that are far from zero. We will use this for our prior distribution of σ except, of course, restricted to the non-negative range, which is ensured from the statement about σ in the parameter block. Hence, our model block in the Stan program will change to the following:

Box 3.12

```
model {
  mu ~ normal(1.589, 0.347);
  sigma ~ cauchy(0,1); //location 0, scale 1
  income ~ lognormal(mu,sigma);
}
```

(Readers need to consult the Stan Reference Manual[3] to see a list of supported distributions and how they are expressed.)

This leads to the following output:

```
Inference for Stan model: model1income.
4 chains, each with iter=4000; warmup=2000; thin=1;
post-warmup draws per chain=2000, total post-warmup draws=8000.

            mean se_mean     sd     2.75%     97.5% n_eff Rhat
mu         10.01    0.00   0.02      9.98     10.05  7208    1
sigma       0.52    0.00   0.01      0.50      0.55  5955    1
mn      25593.62    5.24 444.69  24752.35  26483.90  7201    1
sn      14353.43    6.32 501.24  13442.65  15373.95  6293    1
```

[3]https://mc-stan.org/docs/2_18/reference-manual/

```
Samples were drawn using NUTS(diag_e) at Sat Feb 16 12:09:50 2019.
For each parameter, n_eff is a crude measure of effective sample
size, and Rhat is the potential scale reduction factor on split
chains (at convergence, Rhat=1).
```

As can be seen, this is almost identical to the previous results when improper priors were used. This only emphasizes that we have a large sample size, and that the results are hardly influenced by the priors. Nevertheless, our point was to show how to introduce the prior distributions into the program.

Using Stan for Inference About a Proportion

Let's return now to the first problem we tackled with Bayesian analysis, which was to estimate the proportion below the poverty line. Recall that X is a random variable denoting the number of people out of $n = 1000$ below the poverty line. This has a binomial distribution with parameters θ, the unobservable probability of being below the poverty line, and $n = 1000$. We observed that $X = 415$. The Stan program to represent this is as follows (the file name is 'binomialmodel.stan'):

Box 3.13

```
data{
  int<lower=0> n;
  int<lower=0,upper=n> x;
}

parameters{
  real<lower=0,upper=1> theta;
}

model{
  theta ~ uniform(0,1);
  x ~ binomial(n,theta);
}
```

This is invoked by

Box 3.14

```
fit <- stan (file = "binomialmodel.stan", # Stan program
        data = binomialdata, # named list of data
```

```
          chains = 4, # number of Markov chains
          iter = 4000, # total number of iterations per chain
          cores = 4, # number of cores
)
```

The results table is obtained as follows:

Box 3.15

```
print(fit, pars=c("theta"), probs=c(.0275,.975))
```

```
Inference for Stan model: binomialmodel.4 chains, each with
iter=4000; warmup=2000; thin=1; post-warmup draws per chain=2000,
total post-warmup draws=8000.

      mean se_mean   sd  2.75% 97.5% n_eff Rhat
theta 0.42   0      0.02 0.38  0.45  2984   1

Samples were drawn using NUTS(diag_e) at Mon Feb 18 16:06:55 2019.
For each parameter, n_eff is a crude measure of effective sample
size, and Rhat is the potential scale reduction factor on split
chains (at convergence, Rhat=1).
```

Note that the mean estimate of θ is shown as 0.42 (the true expected value was computed earlier as 0.4152). Also the 95% credible interval is almost identical to the true one computed earlier.

Let's consider the highest density interval instead of the equal tails credible interval. In order to be able to do this, we need to install the R package 'bayestestR' (Makowski et al., 2019). Please see the examples indicated in the Online Resources section. Using this package, finding an HDR is very easy.

Box 3.16

```
e <- extract(fit)
hdi(e$theta, ci = 0.95)
```

First, we extract information from the fit as before ('e') and then call the 'hdi' function on the posterior distribution values stored in 'e$theta', and then specify the credible interval ('ci') that we require – in this case, 95%.

The result is [0.39, 0.45]. In this case, it is almost identical to the equal tails interval. This is because the posterior distribution is almost symmetric.

The general lesson is that although sometimes it is possible to work through the mathematics and compute explicit algebraic expressions for posterior distributions, mostly this is not possible. Using a program such as Stan is much easier, and of course, although carried out through numerical integration, it will give very good results most of the time.

de Finetti's Exchangeability Theorem

This section is of theoretical interest. It provides a powerful mathematical justification for the Bayesian approach to statistical inference. It is about a theorem due to Bruno de Finetti (Cifarelli and Regazzini, 1996; Regazzini and Bassetti, 2008). The starting point is the idea of exchangeability. Suppose that $X_1, X_2, X_3, \ldots$ is an infinite sequence of random variables. Consider any n of these, with joint pdf $f(x_1, x_2, \ldots, x_n)$. The sequence is exchangeable if for any permutation of the indices 1,2, ..., n the joint probability remains unchanged, so as *one particular example*,

$$f(x_1, x_2, \ldots, x_n) = f(x_n, x_{n-1}, \ldots, x_1)$$

and this holds for all $n!$ permutations of the indices, for any subset of the sequence for any n.

If the X_i are independent, then obviously they are also exchangeable. However, a sequence can be exchangeable without being independent. For example, consider the sequence $Z_i = Y + X_i$, where Y is a random variable independent of the X_i. The random variables in this sequence are not independent since, for example, knowing something about Z_k tells us something about Z_m for any k and m, because they are both dependent on Y. However, they are exchangeable because the order in which they are specified in the joint distribution makes no difference.

de Finetti's theorem shows that a sequence is exchangeable if and only if there exists a parameter θ such that the sequence is independent conditional on this, and the joint probability of the sequence can be written as

$$f(x_1, x_2, \ldots, x_n) = \int [f(x_1|\theta)\, f(x_2|\theta) \ldots f(x_n|\theta)]\, p(\theta) d\theta \tag{3.6}$$

where $p(\theta)$ is a probability density for θ. Notice that conditional independence of the x_i given θ is expressed in the square bracket of Equation (3.6). In fact, if we write $\mathbf{x} = (x_1, x_2, \ldots x_n)$, then

$$f(\mathbf{x}|\theta) = f(x_1|\theta)\, f(x_2|\theta) \ldots f(x_n|\theta)$$

is the likelihood of the set of observations, and

$$f(\mathbf{x}|\theta)\, p(\theta) = f(\mathbf{x}, \theta)$$

which is the joint distribution of $\mathbf{x}$ and θ. Hence, integrating out over θ gives the unconditional joint distribution of $\mathbf{x}$.

What this theorem is showing us is that for exchangeable variables there must be a parameter θ, which has a probability distribution, and is such that the variables are conditionally independent given θ. In other words, there is a likelihood, and a probability distribution over a parameter, which can be thought of as a prior. This parameter must exist, and Equation (3.6) must hold. This is essentially a mathematical proof of the ideas behind Bayesian statistics.

Summary

This chapter is the core of this book. Understanding the concepts and methods (if not the mathematical details) presented here is critical. We started with a simple numerical example and showed how Bayes' theorem can be used to solve problems involving conditional probabilities. We then showed, with a simple change of meaning (D = doctor to D = observable data; W = woman to W = statement about the world), how Bayes' theorem can be used in statistical inference. It is used to update prior probabilities about unobservables in the light of observed data. It is based on the critical components of the likelihood (a model of how the data vary conditional on the unobservable parameters), the prior distributions of the parameters, and then the derived posterior distribution.

We showed how this method can be applied to estimating the proportion below the poverty line, demonstrating the mathematical derivation of the posterior distribution. We also showed how prior information can be taken into account (e.g. to consider the extent to which the strongly held views of the Borgonian minister would be supported). We then looked at the more complex problem of estimating the mean income, first using a clearly incorrect likelihood, representing the distribution of income as normal. We used this to derive the Bayesian interpretation of a classical statistical test (Student's t test), and we saw that this is equivalent to the assumption of no prior information. We also showed that the predicted distribution of income from this model was not adequate. We then switched to a more appropriate lognormal distribution and showed how this solution could be represented as a program using the Stan language. We showed how to introduce prior distributions. We finally showed how to solve the originally presented problem of estimating the number below the poverty line, also using a Stan program. Finally, we introduced some theory that shows some of the underlying mathematical framework of the Bayesian approach to statistical inference with de Finetti's theorem.

In the next chapters, we will simply amplify the ideas presented here with different situations and different examples. However, no new fundamental concepts will be introduced – the rest is application.

Online Resources

The R examples that go along with this chapter can be found at

www.kaggle.com/melslater/slater-bayesian-statistics-3

Four

RELATIONSHIP

Introduction

In the previous chapter, we looked at an example involving the distribution of income. Apart from obvious matters such as the standard of living of a family, what else might income influence? A hotly debated topic in Borgonia concerns the relationship between income and health. The Borgonian authorities claim the best health service in the world, and while they accept as a 'natural' that there is inequality of income and the consequences of this (e.g. some families can afford expensive houses, cars, and holidays), they do not accept that health is impacted, because the health service acts as a 'leveller'. However, others argue that notwithstanding the health service, basic inequalities are detrimental to the health of those at the lower end of the income range. The strong negative relationship between income and health is well known in other countries – for example, Chetty et al. (2016) looked at the relationship between predicted age at death of people aged 40 and their income and found the greater the income, the greater the estimated age at death.

A survey in Borgonia from official government statistics was obtained of 200 self-identified men and 200 self-identified women recording their age at death and their estimated annual household income. The survey also found that approximately 0.85% of the population identified as non-binary. However, the scope of the study was on age of death of self-identified women and men, as further effort is being put into a much larger scale analysis across a range of genders. The study has to be larger scale in order to obtain a representative sample. Methods of data collection are beyond the scope of this book, but for an example of research estimating the non-binary population, see Spizzirri et al. (2021). We will continue to use the terms 'men/male' and 'women/female' as labels for self-identified and refer to this as 'sex' rather than the broader category of gender. For women and men, the relationship between age of death and income in this sample is shown in Figure 4.1.

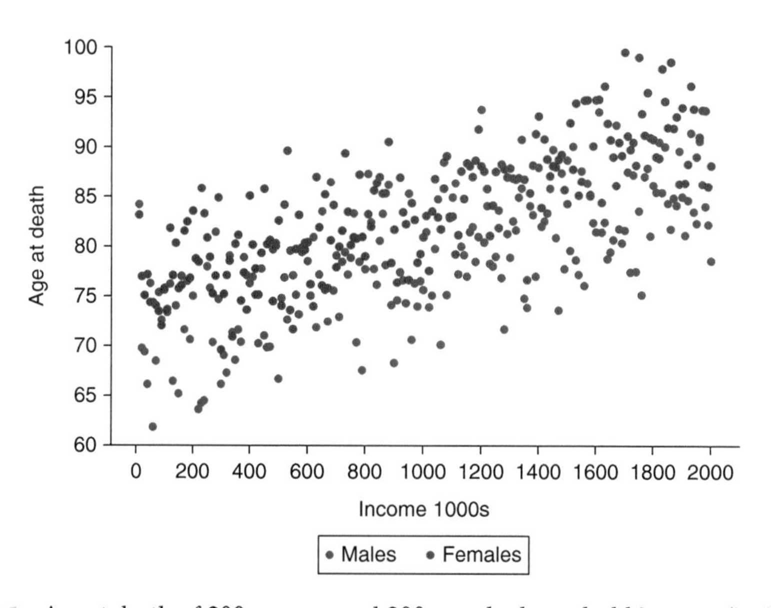

Figure 4.1 Age at death of 200 women and 200 men by household income (in 1000s b€)

It seems that there is a strong positive linear relationship between income and age at death, and moreover it seems different for men and woman. Of course, the relationship is not an exact one. How can we model this? *Consider the data for the men only for the moment.* Let's refer to income as the variable x and age at death as the variable y. Then it is reasonable to suppose that $y = \beta_0 + \beta_1 x$, where β_0, β_1 are unobservable parameters. However, looking at Figure 4.1, we know that this relationship is wrong, for it implies an exact straight line. We could reformulate it as

$$E(y|\beta_0, \beta_1, x) = \mu = \beta_0 + \beta_1 x$$

Here we postulate that y has a probability distribution with mean given by the expression above. For any fixed value of β_0, β_1, x, the *mean* income, is given by the expression. Since this is the mean, we would expect random variation about it according to the probability distribution of y. Although we have no evidence one way or the other for this, we will assume that the distribution of y is normal, with unknown standard deviation σ, and see how that goes. Our model is now

$$(y|\beta_0, \beta_1, x) \sim normal(\beta_0 + \beta_1 x, \sigma) \tag{4.1}$$

β_0, β_1 and σ are the unknown parameters (the unobservables), and there is a sample (x_i, y_i), $i = 1, 2, \ldots, n = 200$ (considering men only) providing the observations. The problem is to be able to make probability statements about the unobservables given the data.

Actually there is nothing new about this. Expression (4.1) will allow us to define the likelihood – the distribution of the observations conditional on the unknown parameters. (Here x is not a parameter but is known, it is not a random variable.) If we had prior distributions on the parameters, we could compute their posterior distributions and then make whatever inferences about them that are of interest. Not only that, but given a set of x values, we could use Expression (4.1) to make predictions about the age of death given any particular income (though under important restrictions).

Example of Derivation of Posterior Distributions

Here we outline how the posterior distributions can be derived mathematically. Readers can skip to the next section if the mathematics here is not interesting. Following from Equation (3.5) of Chapter 3, the likelihood is

$$f(y_1, y_2, \ldots, y_n|\beta_0, \beta_1, \sigma) = \left(\frac{1}{\sigma\sqrt{2\pi}}\right)^n \exp\left(-\frac{1}{2\sigma^2}\sum_{i=1}^{n}(y_i - \beta_0 - \beta_1 x_i)^2\right)$$

Starting from prior distributions for the unknowns and using the likelihood, we can go through the mathematics and find expressions for the posterior distributions, as we did in Chapter 3. To illustrate this, we let the prior distributions for $\beta_j \sim normal(0, \sigma_j)$, $j = 0, 1$. We would use this in circumstances where we have little prior information about the parameters and include 0 as possibilities. The values of the σ_j would depend on the scale of units of measurement (we will discuss this in more detail below). If we repeat the analysis carried out in Chapter 3, then, after a lot of tedious mathematical manipulations, we obtain the posterior distribution of β_0 conditional on β_1 and the standard deviations as

$$\beta_0 \mid \beta_1, \sigma, \sigma_0, \sigma_1 \sim normal\left(n\left(\frac{\sigma_0^2(\bar{y} - \beta_1\bar{x})}{\sigma_0^2 n + \sigma^2}\right), \sqrt{\left(\frac{\sigma^2\sigma_0^2}{\sigma_0^2 n + \sigma^2}\right)}\right)$$

where $\bar{x}$ and $\bar{y}$ are the sample means.

We can make the prior distribution for β_0 *improper* by setting $\sigma_0^2 \to \infty$. Then,

$$\beta_0 \mid \beta_1, \sigma \sim normal\left(\bar{y} - \beta_1\bar{x}, \frac{\sigma}{\sqrt{n}}\right)$$

This is equivalent to the result for simple linear regression from frequentist statistics. The mean makes sense because $E(y|\beta_0, \beta_1, \sigma) = \beta_0 + \beta_1 x$, so that $\beta_0 = E(y|\beta_0, \beta_1, \sigma) - \beta_1 x$. Moreover, the standard deviation $\frac{\sigma}{\sqrt{n}}$ is just the standard deviation of $\bar{y}$.

The posterior distribution of β_1 conditional on all the standard deviations is

$$normal\left(\frac{\sigma_1^2\left((\sigma_0^2 n + \sigma^2)\sum x_i y_i - n^2\bar{x}\bar{y}\sigma_0^2\right)}{(\sigma_0^2 n + \sigma^2)(\sigma^2 + \sigma_1^2\sum x_i^2) - \sigma_0^2\sigma_1^2 n^2\bar{x}^2}, \frac{\sigma\sigma_1\sqrt{\sigma_0^2 n + \sigma^2}}{\sqrt{(\sigma_0^2 n + \sigma^2)(\sigma^2 + \sigma_1^2\sum x_i^2) - \sigma_0^2\sigma_1^2 n^2\bar{x}^2}}\right)$$

Don't worry, you won't have to ever deal with this!

Once again, we can make the prior distributions improper by setting $\sigma_0^2, \sigma_1^2 \to \infty$, in which case the posterior distribution becomes

$$\beta_1 \mid \sigma \sim normal\left(\frac{\sum(x_i - \bar{x})(y_i - \bar{y})}{\sum(x_i - \bar{x})^2}, \frac{\sigma}{\sqrt{\sum(x_i - \bar{x})^2}}\right)$$

Again this is equivalent to the classical result from statistics. The mean is the 'least squares estimator' for β_1 and the standard deviation is the corresponding value. All of this can be shown to give rise to Student's *t* tests on the parameters from a classical statistics point of view. This further illustrates that typically results from classical statistics may be obtained from Bayesian analysis using improper priors.

Note that this also provides another example of conjugate prior distributions. The posterior distributions for the β_j are normal, just like the priors, because the normal priors are conjugate to the normal likelihood function.

Estimating the Linear Relationship for Males Only

We have seen above that even for a relatively simple problem the mathematics is tedious. Although we obtained formulae for the posterior distributions, this was for a particular case where conjugacy was possible. In most cases, there is no conjugate prior or the mathematics is anyway daunting, or the resulting integrals to find the posteriors can only be solved numerically. Trying to work out the mathematics is not only tedious, or often impossible, but also unnecessary. Here instead, we show how the problem may be solved using the Stan programming language as in the previous chapter. Consider the situation for the income and age at death data.

Box 4.1

The data:
$n = 200$
y is a vector of length n (the age at death)
x is a vector of length n (the income in 000s)

The parameters
β_0, β_1 and $\sigma > 0$

The model:
prior distributions of β_0, β_1 and $\sigma > 0$ are improper.
$y \sim normal(\beta_0 + \beta_1 x, \sigma)$

This becomes the Stan program:

Box 4.2

```
data {
  int<lower=0> n; //number in sample
  vector<lower=0>[n] x; //income 000s
  vector<lower=0>[n] y; //age at death
}

parameters {
  vector[2] beta;
  real<lower=0> sigma;
}

model {
  for(i in 1:n){ //note that array indexing starts at 1
    y[i] ~ normal(beta[1] + beta[2]*x[i],sigma);
  }
}
```

Running this with 2000 iterations, we obtain the summary of the posterior distributions shown in Table 4.1.

Table 4.1 Summaries of the posterior distributions using improper priors for all parameters

Parameter	Mean	SD	2.5%	97.5%	Effective sample size	Rhat
beta[1] β_0	71.38	0.65	70.10	72.65	1675	1
beta[2] β_1	0.01	0.00	0.01	0.01	1976	1
sigma σ	4.77	0.24	4.31	5.25	1821	1

The R commands to do this are available in the Online Resources section.

The simulation results in 4000 observations on each of the parameters, thus estimating the posterior distributions. These include 'warm-up' (or 'burn-in') observations, which we will discuss in a later chapter. Figure 4.2 shows the corresponding histograms. Also within R, there is a function ('density') that estimates the corresponding density curve based on the same data. These are shown alongside the histograms. The density curves are slightly 'wonky' in places, emphasizing the fact that these are estimates based on a simulation and not based on plots of the true mathematically obtained posterior distributions. As we saw in Chapter 3, these estimates are nevertheless typically very close to the true distributions, provided that the simulation has converged.

We can see, of course, as is obvious from Figure 4.1, that there is a strong linear relationship between age at death and income. The parameter β_0 tells us something about the likely age at death for a hypothetical 0 income, and the parameter β_1 tells us about the proportional increase in age of death with increase in income. It is clear that $\beta_1 > 0$. In fact, the posterior probability $P(\beta > 0|data) = 1$.

In order to obtain posterior predicted distributions for the age at death of each individual, we add the 'generated quantities' block to the Stan program.

Box 4.3

```
generated quantities{
  vector[n] y_new;
  for(i in 1:n){
    y_new[i] = normal_rng(beta[1] + beta[2]*x[i],sigma);
  }
}
```

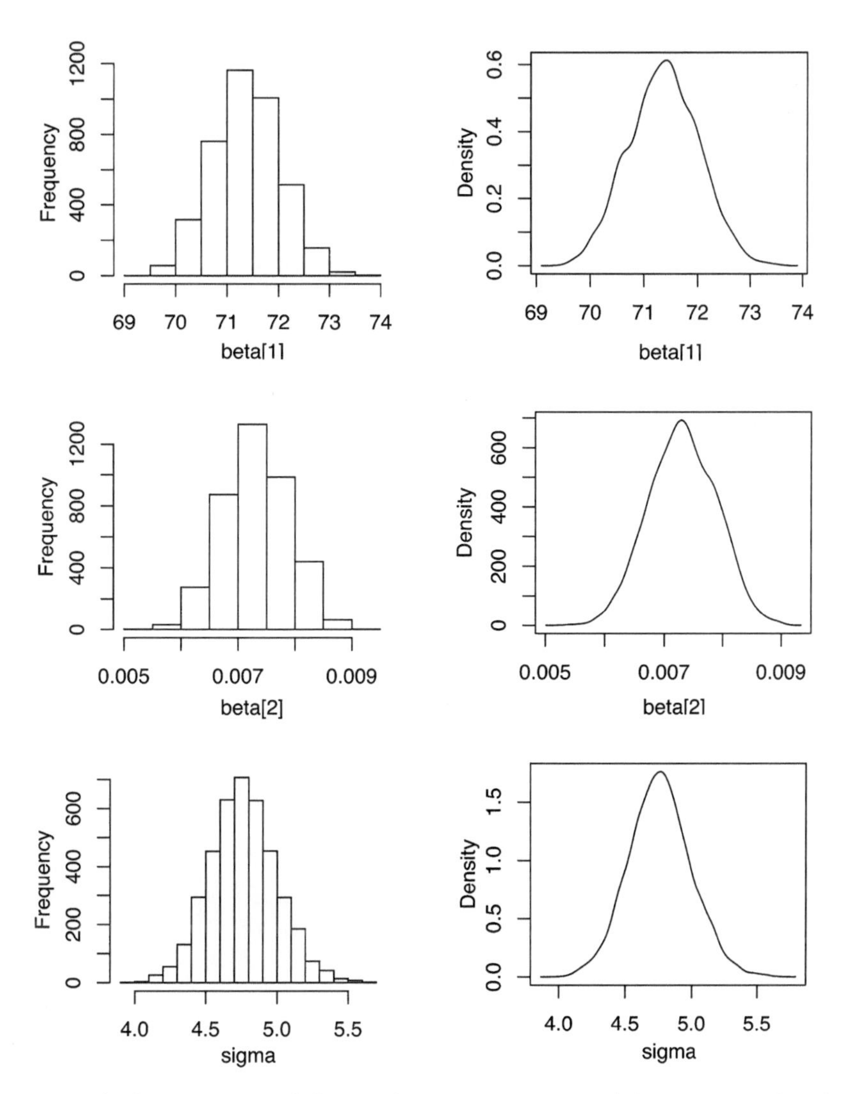

Figure 4.2 The histograms and density function estimates of the posterior distributions of the parameters of the model based on the fit with improper priors

Recall that this simulates observations on the model: 4000 simulated observations on each of the 200 individuals. For example, the probability densities for age of death are shown in Figure 4.3 for two individuals, one with a low income and the other with a high income (at the two extremes of the data). From this, we can compute, for example,

For low income: *P(predicted age of death* > 90|*data*) = 0.00025
For high income: *P(predicted age of death* > 90|*data*) = 0.2015
Quite a difference!

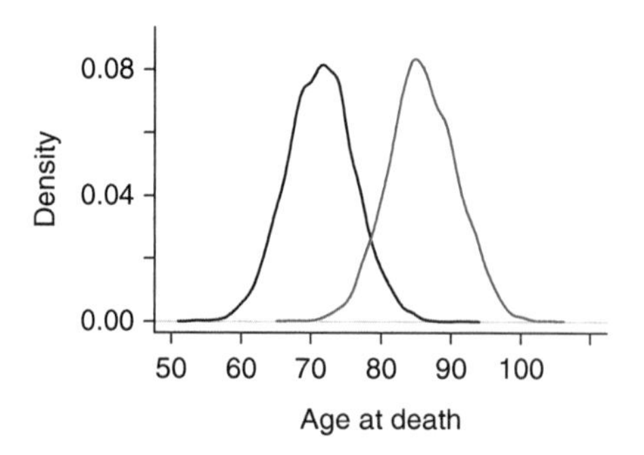

Figure 4.3 Posterior distributions for two individuals – curve in black represents someone who had a low income and blue a high income

For each individual, we can find the mean of the posterior predicted distribution of age at death and use this as a point estimate. Figure 4.4 shows the plot of these point estimates against the observed value for each individual. It is clear that the fit is a good one. The correlation between observed and fitted is $r = 0.67$ ($R^2 = 0.44$).

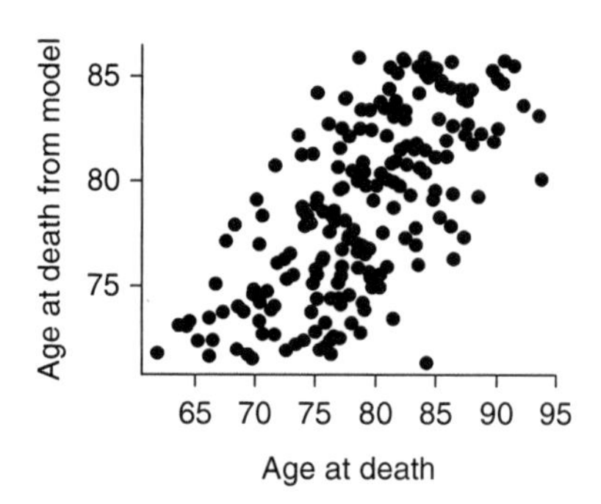

Figure 4.4 Age at death estimated by the means of the posterior distributions of the predicted values for each individual, by the actual age at death

Using More Informed Priors

The analysis above used improper priors – that is, that β_0, β_1 could be any values at all (even negative) and σ could be any positive value. The data set is quite large ($n = 200$), and the reader is invited to try different priors to see what difference this makes. The answer is that the data dominate, and different priors do not make much difference.

However, suppose that $n = 20$ rather than 200. In this case, the priors might make a difference, and we can do better than improper priors. Consider β_0. This represents the age at death when income is zero. Is this likely to be 2,145,678 or –42? Are these two possibilities equally likely? No, of course we know something in advance about β_0. At a minimum, we can adopt the biblical proverb 'three score and ten' (70) as the mean expected age. It is certainly possible to be as high as 80 or as low as 60. This suggests that a reasonable prior for β_0 could be *normal*($\mu = 70$, $\sigma = 5$), which gives approximately a 95% prior probability that $60 < \beta_0 < 80$ (the mean ± 2 standard deviations contains about 95% of the distribution). This is, of course, the *prior* 95% credible interval. We could do even better than this since age also has lower and upper limits. This would lead to the idea of 'censored' distributions – that is, we could cut the normal distribution at a lower and upper limit. For example, in Stan we could specify the prior distribution as

```
beta[1] ~ normal(70,5);
```

Alternately, we could introduce lower and upper limits,

```
beta[1] ~ normal(70,5) T[0,120];
```

which specifies a truncated or censored normal distribution with lower bound 0 and upper bound 120. In practice, in this particular example, it would not make any difference, because 99% of the distribution lies in the interval $\mu \pm 2.58\sigma$, in this case: $[70 - 2.58 \times 5, 70 + 2.58 \times 5] = [57.1, 82.9]$. So the probability of extending to 0 or 120 is negligible. But there are other situations where using censored distributions might be relevant.

Now consider β_1. Here we are not so much in the realm of the obvious, because learning about β_1 is the purpose of the investigation. β_1 represents the relationship between mortality and income. However, there have been previous studies, and we can expect that $\beta_1 > 0$ (i.e. although possible, it is highly unlikely that greater income leads to earlier death – except in some exceptional cases such as people with extreme lifestyles facilitated by their great wealth). β_1 represents the change in age at death for 1000-unit increase in income (a change of 1 in x). Is β_1 likely, for example, to be 4? This would imply that age at death for someone with an income of 200,000 would be $70 + 200 \times 4 = 870$, and for someone with an income of say 2,000,000, it would be $70 + 2000 \times 4 = 4070$. This doesn't really work (and also emphasizes the inadequacy of the improper prior that gives equal weight to these possibilities). Let's work backwards and say that at the highest level of income (say 2,000,000) a high age might be 90. Then $70 + \beta_1 2000 = 90$. This would give $\beta_1 = 0.01$, which is some idea about the order of magnitude of β_1. Let's therefore consider the prior as $\beta_1 \sim$ *normal*(0.01, 0.001). This would give a 95% prior credible interval as $0.008 < \beta_1 < 0.012$, and there is still 5% probability to be outside of this range.

Now let's consider the standard deviation σ. For someone at minimum income, we postulated 70 as estimated age at death. If we choose a standard deviation of around 30, then this gives very wide leeway – between approximately 10 and 130. Since we do not know much about σ except that it is non-negative, we will choose a distribution with very wide support, but which is nevertheless a proper distribution. For example, the Cauchy distribution with parameters 0 and 1.25, restricted to being non-negative, would result in a 95% prior credible interval for σ of approximately 0.05 to 31. We will use this as the prior.

Figure 4.5 shows the differences in posterior distributions depending on whether the improper priors or the priors just introduced were used. For β_0 and β_1, the differences are noticeable, but there is no difference for σ.

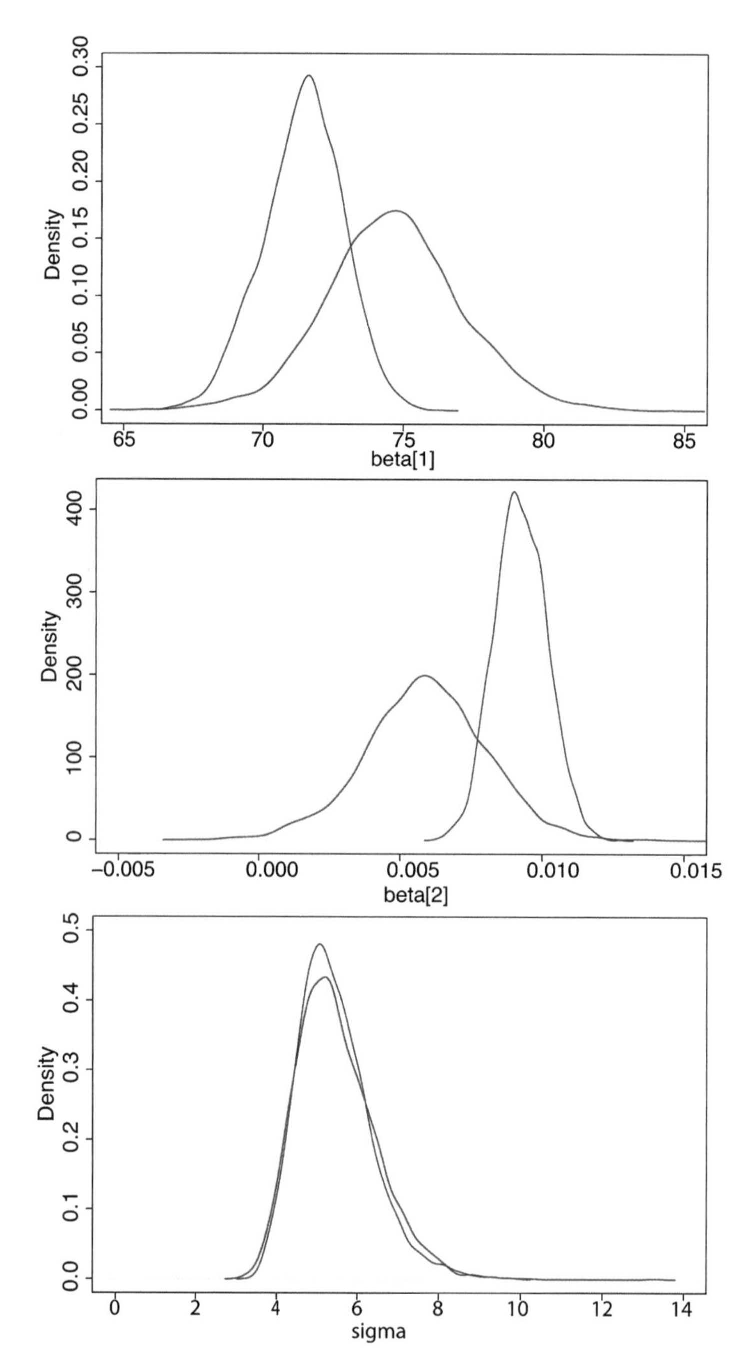

Figure 4.5 Posterior distributions for $n = 20$ observations, comparing the improper prior distribution results (red) with the proper prior distributions (blue)

If we consider the overall goodness of fit, using the means of the posterior predicted ages, as in Figure 4.4, there is not much difference between the two methods based on different priors. However, there are substantial differences with respect to specific probabilities.

Table 4.2 Posterior probabilities of death at age 90 or more by prior distribution and income level

Income level	Improper priors	Proper priors
Low income	0.00825	0.00125
High income	0.24825	0.4315

Table 4.2 shows the posterior probabilities age of death at least at 90. The two methods give substantially different results.

Now it could be argued that the proper priors have built the result into the method. However, the assumptions that went into the construction of these priors are openly stated and open to criticism and revision. As we have seen, with sufficient data the priors don't have a great influence. However, with a relatively small amount of data, here 20 individuals, similar to the number of participants in a typical psychology experiment, the priors can be influential. The point is that they should be argued for, and surely they are better than just postulating that all possible values of the parameters are equally likely. When we have information, or just common sense, we should use it.

Introducing a Factor

The above is based only on the data for males. However, there are also 200 observations on females. We should include these data, especially because this will allow us to address the issue as to whether there is a difference between males and females with respect to age at death in relation to income – as suggested by Figure 4.1. Now we have the following situation:

Box 4.4

The data:

$n = 400$

y is a vector of length n (the age at death)

x is a vector of length n (the income in 000s)

s is a vector of length n with coding 0 = Male, 1 = Female with 200 observations on each.

(Continued)

The parameters
$\beta_0, \beta_1, \beta_2$ and $\sigma > 0$

The model:
prior distributions of $\beta_0, \beta_1, \beta_2$ and $\sigma > 0$ are proper.
$y \sim normal(\beta_0 + \beta_1 x + \beta_2 s, \sigma)$

The variable 'sex', which we denote as s, is a *factor*. A factor is a type of variable which takes a value from a set of categories, here (Male, Female) (coded as 0 or 1, respectively). In this case, the factor is binary, it has two *levels*, but in general, the number of levels could be any $k > 1$. For example, suppose that we had included not 'sex' but 'gender', for example only ('Male', 'Female', 'Non-Binary', 'Prefer not to say'). This factor has four levels.

The case of a binary factor (i.e. with two possible levels) is particularly easy to introduce into the model and understand – since it is simply 0 or 1. Hence β_2 represents the amount by which age at death may be different between females and males.

In the case of a factor with four levels, we would need three 0/1 variables to account for this: for example, genderM = 1 for male and 0 otherwise, genderF = 1 for female and 0 otherwise, genderB = 1 for binary and 0 otherwise. We do not need a fourth variable, since once the first three are given, the last one is specified. For example, if genderM = genderF = genderB = 0, then it must be the case that 'prefer not to say' is 1. In general, if a factor has k levels, then $k - 1$ binary variables are needed in order to be able to include it in a model. These are usually called 'dummy variables'.

Box 4.5

```
data {
  int<lower=0> n; //number in sample
  vector<lower=0>[n] x; //income 000s
  vector<lower=0>[n] y; //age at death
  vector<lower=0,upper=1>[n] s; //male = 0, female = 1
}

parameters {
  vector[3] beta;
  real<lower=0> sigma;
}

model {
  beta[1] ~ normal(70,5);
```

```
  beta[2] ~ normal(0.01,0.001);
  beta[3] ~ normal(0,5);
  sigma ~ cauchy(0,1.25);
  for(i in 1:n){ //note that array indexing starts at 1
    y[i] ~ normal(beta[1] + beta[2]*x[i] + beta[3]*s[i],sigma);
  }
}

generated quantities{
  vector[n] y_new;
  for(i in 1:n){
    y_new[i] = normal_rng(beta[1] + beta[2]*x[i]  +
                          beta[3]*s[i],sigma);
  }
}
```

The program that specifies the model is similar to the first one, except with the introduction of the additional parameter ('beta[3]' representing β_2). We choose as the prior for this *normal*(0,5), meaning that we have a prior 95% probability that the female age at death is within ±10 years.

Table 4.3 Summaries of the posterior distributions of the model including sex

Parameter	Mean	SD	2.5%	97.5%	Effective sample size	Rhat
beta[1] β_0	70.45	0.46	69.54	71.35	2086	1
beta[2] β_1	0.01	0.00	0.01	0.01	2890	1
beta[3] β_2	5.55	0.43	4.72	6.40	1902	1
sigma σ	4.34	0.15	4.05	4.65	2291	1

Table 4.3 summarizes the posterior distributions. β_0, β_1, and σ are very similar to the results under the earlier analysis. The 95% credible interval for β_2 is clearly way above 0; in other words, there is a substantial difference between men and women with respect to age of death – around 5½ years in favour of women.

The overall model fit is good as shown by Figure 4.6. The correlation between the observed and simulated values is $r = 0.78$ ($n = 400$; $R^2 = 0.61$).

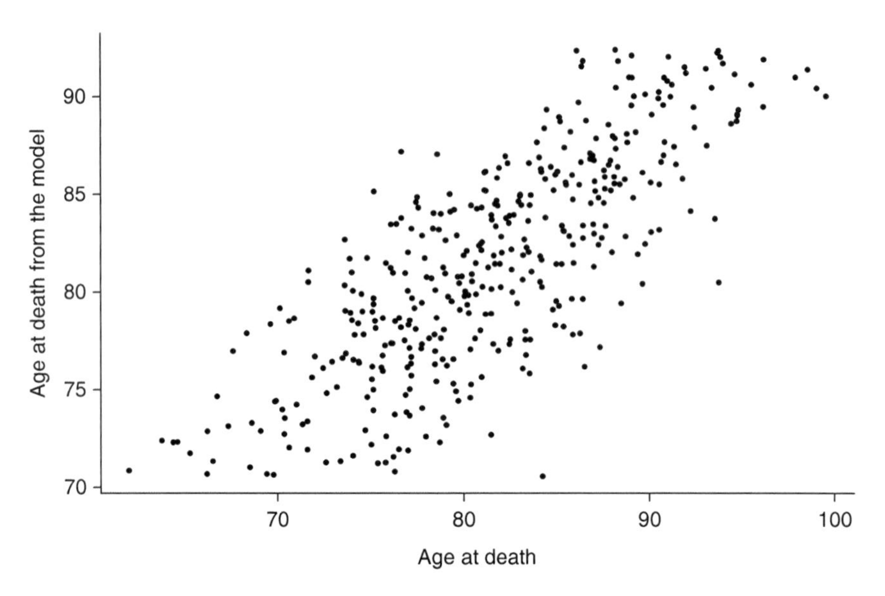

Figure 4.6 Age at death estimated by the means of the posterior distributions of the predicted values for each individual, by the actual age at death, for the full data set including sex

Adding an Interaction Term

The overall findings then are that there is a positive relationship between income and age at death and that irrespective of the level of income, women on the average lived about 5½ years longer than men. However, it is possible that the level of income might impact men and women differentially. At the moment, our model is essentially two parallel straight lines (of course, with random variations about those lines) – where the lower line represents the relationship between income and age for men, and the upper one for women. But perhaps it isn't like that. Our model does not allow for the possibility that the lines are not parallel, and that the difference between the ages of death for women and men is also influenced by income. We can overcome this limitation by adding a further parameter β_3, which is the coefficient of the product $(s_i \times x_i)$. If $\beta_3 \neq 0$, then the slope of the theoretical line relating age at death (y) to income (x) can be different for males and females. This term is called an *interaction*. It allows for interaction between sex and income that contributes above and beyond how they individually contribute to age and death. Interaction terms represent the idea that 'the whole is more than the sum of the parts'.

The new situation is shown below, where there are changes in the parameters, model, and generated quantities blocks, as shown in the red highlights below.

Box 4.6

```
data {
  int<lower=0> n; //number in sample
  vector<lower=0>[n] x; //income 000s
  vector<lower=0>[n] y; //age at death
  vector<lower=0,upper=1>[n] s; //male = 0, female = 1
}

parameters {
  vector[4] beta;
  real<lower=0> sigma;
}

model {
  beta[1] ~ normal(70,5);
  beta[2] ~ normal(0.01,0.001);
  beta[3] ~ normal(0,5);
  beta[4] ~ normal(0,10);
  sigma ~ cauchy(0,1.25);
  for(i in 1:n){ //note that array indexing starts at 1
    y[i] ~ normal(beta[1] + beta[2]*x[i] + beta[3]*s[i] +
                  beta[4]*(s[i]*x[i]),sigma);
  }
}

generated quantities{
  vector[n] y_new;
  for(i in 1:n){
    y_new[i] = normal_rng(beta[1] + beta[2]*x[i]  + beta[3]*s[i] +
                          beta[4]*(s[i]*x[i]),sigma);
  }

}
```

We do not have any prior insight or knowledge about the possible values of β_3; therefore, we give it a wide support prior distribution. The prior 95% credible interval is approximately –20 to 20. Of course, the extremes are highly unlikely. Nevertheless, it is preferable to give such a weak prior than an improper prior which gives equal weight to all possible values.

Table 4.4 Summaries of the posterior distributions of the model including sex and an interaction term

Parameter	Mean	SD	2.5%	97.5%	Effective sample size	Rhat
beta[1] β_0	70.857	0.546	69.829	71.949	1626	1.003
beta[2] β_1	0.008	0.000	0.007	0.009	1975	1.002
beta[3] β_2	4.657	0.799	3.126	6.226	1734	1.003
beta[4] β_3	0.001	0.001	0.000	0.002	2089	1.001
sigma σ	4.317	0.154	4.019	4.633	1776	1.002

Table 4.4 summarizes the posterior distributions. As we can see, the mean of the distribution of β_2 has decreased a bit, and the 95% credible interval for β_3 has approximately zero (to three decimal places) at its lower left. The posterior density is shown in Figure 4.7. Also examine the posterior credible interval, which is extremely narrow compared to the prior.

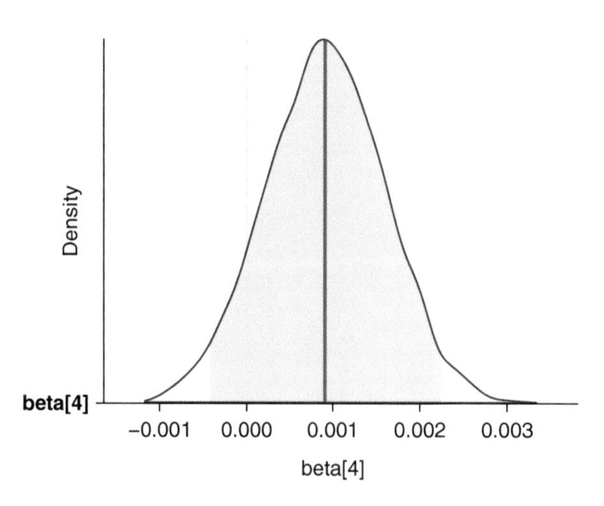

Figure 4.7 Posterior density for β_3. The vertical line is the mean and the shaded area is the 95% credible interval

From the posterior distribution, $P(\beta_3 > 0|Data) = 0.905$. Hence, the probability is that there is an interaction between sex and income in their association with age at death. However, the effect, if there is one, is small. Compare the mean parameter estimates of β_2 (4.7) with β_3 (0.001), so that the interaction term has a far smaller effect than the main term relating income to age. As can be seen from Table 4.5, the inclusion of the interaction term does make some difference to the prediction of age at death after 90 for the higher income group. It is important to realize that although the probability for a parameter being non-zero may be high, care should be taken to check the mean value of the coefficient (or some other estimate of its likely value). If this is very small, then the variable of which the parameter is a coefficient (in this case, the interaction term) may have some explanatory value for the response variable (y, age at death), but that value may be negligible. This is referred to as *effect size*.

Table 4.5 Posterior probabilities of death at age 90 or more by inclusion of interaction term or not

Income level	No interaction term	With interaction term
Low income	0.0	0.0
High income	0.247	0.213

Bayes Factors

The method we have adopted in order to make inferences about unobservable parameters is to compute their posterior distributions and then use these to compute any probabilities in which we are interested. Sometimes, there may be a particular range of values of interest or specific value, and the question is whether the data support this or alternatively some other range or value. In classical statistics, this is referred to as 'hypothesis testing' – typically let the data choose between two or more alternatives, or more accurately, choose whether to reject one of the alternatives (usually called the 'null hypothesis') in favour of the other.

For example, suppose the hypothesis was that the ages of death of men and women were within ±5 years of one another at any level of income. Let's consider the model without the interaction term, recalling that our model is

$$y \sim normal(\beta_0 + \beta_1 x + \beta_2 s, \sigma)$$

The hypothesis is that $\beta_2 \in [-5, 5]$. Let's refer to this hypothesis as H_0, the 'null hypothesis'. Let's consider the 'alternative hypothesis' H_1 as $\beta_2 > 5$: that is, that the difference in age at death for any level of income is at least 5 years. In general, we can consider

$$\frac{P(H_1 \mid Data)}{P(H_0 \mid Data)} = \frac{P(Data \mid H_1)P(H_1)}{P(Data \mid H_0)P(H_0)}$$

$$= \left(\frac{P(Data \mid H_1)}{P(Data \mid H_0)}\right) \times \left(\frac{P(H_1)}{P(H_0)}\right)$$

$$posterior\ odds = likelihood\ ratio \times prior\ odds$$

The likelihood ratio is referred to as the *Bayes factor* in favour of H_1 compared to H_0:

$$BF_{10} = \frac{P(Data \mid H_1)}{P(Data \mid H_0)}$$

It is the multiplier that converts the prior odds into the posterior odds. The greater the value, the more the results are in favour of H_1 compared to H_0. Of course, the reciprocal BF_{01} is the evidence favouring H_0 compared to H_1. The idea of Bayes factors is usually attributed to a 1936 paper by Harold Jeffreys (see, e.g., Etz and Wagenmakers, 2017).

Interpretation of Bayes factors is somewhat arbitrary. Based on original suggestions of Jeffreys (1998), common criteria are that evidence in favour of H_1 is 'anecdotal' for BF_{10} between 1 and 3, 'moderate' up to 10, 'strong' up to 30, 'very strong' up to 100, and 'extreme' thereafter. (Of course, taking the reciprocal gives the same suggestions for H_0.) Various authors have put forward slightly different schemes.

Bayes Factor for an Interval Hypothesis

Now let's return to the particular example of the age at death data:

$$H_0: \beta_2 \in [-5, 5]$$

$$H_1: \beta_2 > 5$$

Recall that the prior distribution is $\beta_2 \sim normal(0, 5)$. The R function 'pnorm(x,mean,sd)' computes the probability that a normal random variable with the given mean and standard deviation (sd) will have a value less than the specified x. Since the distribution is symmetric around 0, we have

$$P(H_1) = pnorm(-5, 0, 5) = 0.1587$$

$$P(H_0) = 1 - 2 \times pnorm(-5, 0, 5) = 0.6827$$

Therefore,

$$prior\ odds = \frac{P(H_1)}{P(H_0)} = 0.2324$$

From the posterior distribution of β_2 ('beta[3]' from the Stan output), we can similarly find that

$$posterior\ odds = \frac{P(H_1 \mid Data)}{P(H_0 \mid Data)} = 7.7336$$

This is straightforward because, as in Chapter 3, we can obtain the posterior distribution of 'beta[3]':

Box 4.7

```
e <- extract(fit)
postHP0 <- mean(e$beta[,3]>-5 & e$beta[,3]<5)
postHP1 <- mean(e$beta[,3] > 5)
postR <- postHP1/postHP0
BF <- postR/priorR
```

This arrives at $BF_{10} = 33.3$. According to the suggested interpretation, this is just over the border between 'strong' and 'very strong' evidence in favour of H_1. This calculation can be seen in context in the Online Resources section.

Bayes Factor for a Specific Value

Now consider the model where we included the interaction term:

$$y \sim normal(\beta_0 + \beta_1 x + \beta_2 s + \beta_3 (s \times x), \sigma)$$

Here we would like to test the hypothesis that $H_0: \beta_3 = 0$ against the alternative that $H_1: \beta_3 \neq 0$; in other words, whether or not the interaction term should be included. The prior is

$$\beta_3 \sim normal(0, 10)$$

(Recall that β_3 is represented by 'beta[4]' in the Stan program.)

Now we cannot proceed as before, since, for example, $P(\beta_3 = 0) = 0$. It would be the same for any specific value. The parameter β_3 is a continuous random variable, so the probability that it takes exactly some particular value is always 0. We could, of course, consider a 'small interval' around 0 and follow the same procedure as above. However, we can employ what is known as the *Savage–Dickey ratio*, which gives an alternate method for computing the Bayes factor in this situation (see Wagenmakers et al., 2010). The ratio is computed as

$$\frac{f_{posterior}(0 \mid H_1, Data)}{f_{prior}(0 \mid H_1)} = \frac{posterior\ pdf\ at\ 0}{prior\ pdf\ at\ 0}$$

This is simply the posterior density function $f_{posterior}$ evaluated at the particular value of interest (here $\beta_3 = 0$) over the prior density f_{prior} function at the same value.

To see how to do this in context using R and the output from Stan, the reader is referred to the Online Resources. We show how to do this here as well. Recall that the posterior distribution for β_3 can be obtained from

```
eI <- extract(fitI)
```

where 'fitI' is the result of the Stan fit, and 'eI$beta[,4]' will be the posterior distribution. However, this is just a sample from the distribution, not an actual function that we can evaluate at a point. In order to represent this as a function, we need to fit a density function to it that can be evaluated at any point (in this case, at 0). For that, we import the 'polspline' library into R and make use of the logspline function[1] (Stone, 1994) as follows:

```
fI <- logspline(eI$beta[,4])

#evaluate at 0

fI0 <- dlogspline(0,fI)
```

[1]https://rdrr.io/cran/polspline/man/logspline.html#heading-5

Here 'logspline' fits a function to the raw distribution data, and 'dlogspline' evaluates this at the particular point 0. Since our prior distribution is *normal*(0,10), we can use the R function 'dnorm(x,mean,sd)' to evaluate the normal distribution with mean and standard deviation (sd) at the value x. Hence, the Bayes factor (Savage–Dickey ratio) is

```
fI0/dnorm(0,0,10)
```

which is a value something over 6000. According to the criteria above, this would be evidence extremely in favour of H_1.

The Bayes Factor Is Highly Influenced by the Prior

The above is not an open-and-shut case in favour of H_1. As we have seen, the data overwhelm the influence of the prior for reasonably sized data sets – especially in our case, the posterior distribution with 200 observations on each of males and females, so that the posterior distribution is not sensitive to the prior. In the Bayes factor, however, since the numerator is dominated by the posterior, and the posterior is not sensitive to the prior, if we change the prior, the numerator does not change, but the denominator does change and, therefore, can result in quite different Bayes factor results. Looking at Table 4.4, it can be seen that the scale of β_3 is very small, the 95% credible interval ranging from 0.000 to 0.002. Therefore the prior, which has a 95% credible interval of approximately –20 to 20, is massively 'inappropriate' – its scale is orders of magnitude greater than the actual scale of β_3. Hence, the denominator in the Savage–Dickey ratio is extremely small. What happens if we change the prior? If the prior becomes, for example, $\beta_3 \sim$ *normal*(0, 0.01), then the prior 95% credible interval is –0.02 to 0.02. This is still much larger than the actual scale. If the analysis is repeated with this prior, then the posterior distribution is hardly affected at all (it is not noticeable to the decimal places in the output), but the revised Bayes factor is 6.3, approximately 1000 times smaller. This is a simple illustration of how sensitive the Bayes factor is to the prior distribution.

We have included the Bayes factor in this book because readers are likely to come across it in the literature or may be advised to use it by others who insist on a 'hypothesis test' more akin to frequentist statistics. However, using it has clear pitfalls, and following through the example, the justification for including β_3 in the model is slight, as we will see in the next section.

Comparing Models

We now have two alternative models for the variation in age at death in relation to income and sex: the one without and the one with the interaction term. As we have seen, the interaction term may contribute. But is it worth keeping interaction term in the model? The Bayes factor analysis gave an inconclusive answer. In general, adding new variables to a model (in this case, the interaction term) *will always improve the model fit*. As the number of explanatory variables approaches the number of observations, so the fit (i.e. the estimation of the response variable values from the explanatory variables) will improve, until in the limit the fit will be exact. So, the specification of a model must always balance between parsimony (the least complexity, the least number of unobservable parameters the better) and goodness of fit. If a model with 10 parameters has just slightly better goodness of fit than a model with 3 parameters, then the smaller model is preferred. The larger model, although having a better fit, might just be modelling noise in the data.

The question is – what is a 'better' or 'worse' goodness of fit of a model?

Let's digress for the moment. Suppose we have n independent trials (e.g. throwing a die), and there are two outcomes on each trial, 'S' (e.g. get a 6) or 'F' (do not get a 6). In the n trials, we observe y Ss. The probability of 'S' is the unobservable θ. We know that the likelihood (probability of the observed data conditional on the unobservable parameters) for this situation is given by the binomial distribution:

$$p(y|\theta) = \binom{n}{y} \theta^{y}(1-\theta)^{n-y}$$

Outside of the Bayesian framework, how could we use this to find an estimate for θ? Since *the result y has actually been observed*, θ should be chosen so that the likelihood is high. Since y is a known value, representing an event that has already happened, it would make no sense for the value of θ to be chosen so that the probability of y occurring is small, because it has occurred. Hence, larger values of the likelihood mean that the model is more plausible than smaller values (having observed the data). This leads to the principle of *maximum likelihood estimation*. In this example, we want to choose θ that maximizes the likelihood. In order to do this, we can use simple calculus. But the expression above is slightly complex. Often, it is simpler to maximize the log of the likelihood (the log likelihood):

$$\log p(y|\theta) = \log\binom{n}{y} + y\log(\theta) + (n-y)\log(1-\theta)$$

In finding the maximum for θ, we would need to differentiate this expression and set the result to 0 and solve for θ. Hence, the terms that do not involve θ at all are irrelevant to this (since they would be 0 in the differentiation). Carrying this out, we find the maximum likelihood estimator for θ to be

$$\hat{\theta} = \frac{y}{n}$$

as would be expected.

A more complex case would be observations y_i ($i = 1, 2, \ldots, n$) on the *normal*(μ, σ) distribution and the problem of estimating the mean and standard deviation. In this case, the likelihood is

$$f(\boldsymbol{y} \mid \mu, \sigma) = \prod_{i=1}^{n} \left(\frac{1}{\sigma\sqrt{2\pi}}\right) e^{-\frac{1}{2}\left(\frac{y_i-\mu}{\sigma}\right)^2}$$

where $\boldsymbol{y}$ is the vector of observations. Since the $\boldsymbol{y}$ have been actually observed, we want the estimates for the mean and standard deviation to be such that the likelihood is high. Using the maximum likelihood principle, we would need to find μ and σ to maximize the likelihood. This is a difficult expression, but taking the log (ignoring parts that do not include the parameters, C)

$$\log f(y \mid \mu, \sigma) = C - n\log(\sigma) - \frac{1}{2\sigma}\sum_{i=1}^{n}(y_i - \mu)^2$$

Then if we differentiate by each of μ and σ and solve the resulting equations, we find that the estimator for μ is the mean $(\bar{y})$ of the y_i and for σ^2, $s^2 = \frac{1}{n}\sum_{i=1}^{n}(y_i - \bar{y})^2$.

The reason for this digression is to emphasize the importance of the log likelihood in assessing the goodness of fit of the model. Conditional on the parameters, the likelihood (and therefore the log likelihood) should be high for the observed data. Recalling the above, we can always increase its value by including more parameters – but at the cost of making the model more complex and increasing the possibility of modelling noise.

Another problem in assessing the suitability of a model is its generalizability. The model has been fitted based on a specific set of data ('in-sample'). Hence, any measure of how good the model is based on in-sample data is going to be an overestimate – since the fitted model is being judged based on the very data that were used to construct it. This is the case for the correlation measure used above. What would be more interesting is to test the model against 'out-of-sample' data – how well can it predict future observations? However, we typically do not have out-of-sample data (and probably, if we did, we would bring them to be in-sample) – at least not immediately to hand.

There are two measures typically used in Bayesian analysis known as *WAIC* (Watanabe–Akaike or widely applicable information criterion) and *loo* (leave-one-out), which are based on log likelihood and try to address the problem (for a full account, see Gelman, Hwang, and Vehtari, 2014; Vehtari et al., 2017; Watanabe, 2010). They are based on the log likelihood of the posterior predicted distribution conditional on the observed data. Suppose $\tilde{y}$ is an out-of-sample observation; then its posterior distribution is $f(\tilde{y}_i|y) = \int f(\tilde{y}_i|\theta)p_{post}(\theta)d\theta$ following the notation of Vehtari et al. (2017). Here $p_{post}(\theta)$ is the posterior distribution of the parameters θ, and y is the vector of original observations. Similarly, we will use $f_{post}(\tilde{y}_i)$ for the posterior predicted density of $\tilde{y}$ given the observations, that is $f(\tilde{y}_i|y)$. This distribution is what can be obtained from the 'generated quantities' block in Stan.

The theoretical quantity of interest is the *expected log pointwise predictive density* (*elpd*) defined as

$$elpd = \sum_{i=1}^{n} E\left(\log(f_{post}(\tilde{y}_i)\right)$$

where the expected value is over the true distribution of y, which of course is unknown. Hence, the *elpd* is estimated by the *log pointwise predictive density* (*lpd*):

$$lpd = \sum_{i=1}^{n}\log(f_{post}(y_i))$$

which can be estimated from the data (e.g. the 'generated quantities' block in Stan). This estimate is referred to as $\widehat{lpd}$:

$$\widehat{lpd} = \sum_{i=1}^{n} \log\left(\frac{1}{S}\sum_{s=1}^{S} f(y_i \mid \theta_s)\right)$$

where θ_s, $s = 1, 2, ..., S$, are the simulated observations on the parameter θ. Recall that in Stan, we might generate say 2000 observations on the parameter θ in order to obtain an approximation to the posterior distribution of θ. The θ_s are just the individual 2000 values that make up this posterior. Moreover, $f(y_i|\theta_s)$ is a known function (the likelihood, we use it to create the model). Hence, $\widehat{lpd}$ is computable from the data and is itself a random variable.

Watanabe-Akaike or Widely Applicable Information Criterion

WAIC (Watanabe et al., 2001) – in the formulation of Vehtari et al. (2017) – is defined as

$$waic = \widehat{lpd} - \hat{p}_{waic}$$

The first part of this is a direct measure of the goodness of fit of the model based on the posterior log likelihoods, and the second term is a penalty for the complexity of the model. It is the sum over each data point of the variances of the log posterior predicted distributions. $\hat{p}_{waic}$ is referred to as the *estimated effective number of parameters*. For any individual observation, the size of $\log(f_{post}(y_i))$ indicates how good the fit is. However, if the variance of this is also large, this indicates the non-reliability of the log likelihood.

Leave-One-Out

The term *loo* stands for 'leave-one-out'. The idea is to partition the original set of observations y_i, $i = 1, 2, ..., n$, into a training set consisting of all but the ith observation $y_{-i} = (y_1, y_2, ..., y_{i-1}, y_{i+1}, ..., y_n)$ and y_i (out-of-sample), and to do this for every $i = 1, 2, ..., n$. The idea is to predict y_i based on y_{-i}. Hence, 'leave-one-out':

$$loo = \sum_{i=1}^{n} \log(f(y_i \mid y_{-i}))$$

In practice, *loo* is approximated from the simulated distributions that are obtainable from the fit (in Stan). Vehtari et al. (2017) show how *loo* can be approximated. In particular, the method results in a set of values k_i, $i = 1, 2, ..., n$, which are called *Pareto estimates*, which give some information about convergence around individual observations. When in principle each observation is removed from the data and the model used to predict it, then we can assess how well that data point is compatible with the model. The k_i, one for each data point, assess this. If $k_i \le \frac{1}{2}$, this indicates no problem, $\frac{1}{2} < k_i \le 1$ indicates possible large variance, and $k_i > 1$ indicates a possible poor model specification.

In this section, we have tried to give some intuitive feel for the measures of the predictive capability of models (and hence their generalizability) without going into the mathematical details.

Luckily, we do not have to do all this work, since there is an R package 'loo' that can be used with Stan.[2,3,4] This can be used to compute both WAIC and *loo* values.

In order to do this, we need to modify the 'generated quantities' blocks slightly to include the computations of the log likelihoods. Here is the example for the model with the interaction term:

Box 4.8

```
generated quantities{
  vector[n] y_new;
  vector[n] log_lik;
  real eta;

  for(i in 1:n){
    eta = beta[1] + beta[2]*x[i]  + beta[3]*s[i]+
          beta[4]*(s[i]*x[i]);
    y_new[i] = normal_rng(eta,sigma);
    log_lik[i] = normal_lpdf(y[i]|eta,sigma);
  }
}
```

Basically, the new variable 'eta' is introduced just to store the current linear predictor, and then use that twice – once for the predicted values ('y_new') and then for the log likelihoods ('log_lik'). Note that 'normal_lpdf' is a Stan function that delivers the log of the probability density function (pdf) of the normal distribution, evaluated at a particular value ('y[i]') and conditional on the mean (in this case, 'eta') and the standard deviation.

In R, in order to make use of this, we use the following commands:

Box 4.9

```
log_lik_I <- extract_log_lik(fitI, merge_chains = FALSE)
r_eff_I <- relative_eff(exp(log_lik_I))
loo_I <- loo(log_lik_I, r_eff = r_eff_I, cores = 2)
print(loo_I) #will print the values for loo
```

[2] https://mc-stan.org/users/interfaces/loo

[3] https://cran.r-project.org/web/packages/loo/vignettes/loo2-with-rstan.html

[4] http://mc-stan.org/loo/articles/loo2-with-rstan.html

We can do the same for the model with no interactions, assuming that the log likelihood computations have been put into the Stan program in the 'generated quantities' block, and then:

Box 4.10

```
log_lik_noI <- extract_log_lik(fit, merge_chains = FALSE)
r_eff_noI <- relative_eff(exp(log_lik_noI))
loo_noI <- loo(log_lik_noI, r_eff = r_eff_noI, cores = 2)
print(loo_noI)
```

Table 4.6 Comparing (A) the two models individually and (B) the difference between them

(A) Each individual model

	No Interaction		Interaction	
	Estimate	***SE***	**Estimate**	***SE***
elpd_loo	–1155.9	13.7	–1155.6	13.7
p_loo	3.6	0.3	4.8	0.4

(B) Comparison between the models

	elpd_diff	**se_diff**
model2	0.0	0.0
model1	–0.3	1.2

In Table 4.6A, the elpd_loo is the *loo* estimate, and p_loo is the effective number of parameters. As we would expect, the elpd_loo estimate for the model with interaction is very slightly larger than the model without interaction (it has more parameters), but the difference is almost nothing.

The output also says: 'All Pareto k estimates are good (k < 0.5)' and plots of these *k* values can be obtained with 'plot(loo_noI)' and 'plot(loo_I)'.

We can compare the two models with the following:

Box 4.11

```
print(loo_compare(loo_noI,loo_I))
```

which outputs the information in Table 4.6B.

This output always shows the model with the greatest loo_elpd first (in this case, model 2) and then differences from that model. Hence, the first row always contains zeros, since each row is the differences from the 'best' model (i.e. itself). It is obvious from Table 4.6A that the model with interaction (model 2) has a slightly higher elpd_loo. The standard error of the difference is based on pairwise comparisons, data point by data point, between the two models. The function 'loo_compare' can take multiple arguments for the comparison of several models.

Comparing the elpd_diff to its standard error, it is clear that there is no advantage to model 2, so that it is not worth keeping the interaction term, which is a very different result as might have been obtained relying solely on the Bayes factor.

We have discussed only a limited number of model comparison methods. The 'loo' is particularly suited to Stan; however, the reader should be aware that there are other methods, and an excellent review article on this topic is Shiffrin et al. (2008) and Chapter 7 of Gelman, Carlin, et al. (2014).

Vectorization

Stan has efficient vector and matrix representations and operations. Hence, instead of specifying that each individual

y[i] ~ normal(beta[1] + beta[2]*x[i] + beta[3]*s[i],sigma)

we can use the vectorized form:

y ~ normal(X*beta,sigma)

where 'X' is, in general, an $n \times k$ matrix where each column is an explanatory variable or represents a factor, and 'y' is the corresponding response variable vector. If an intercept term is required (β_0 or 'beta[1]' in the program) then the first column of 'X' has to be a column of 1s.

Here is the new program for the case introducing sex as a factor but without the interaction:

Box 4.12

```
data {
    int<lower=0> n;          //number in sample
    int<lower=0> k;          //number of parameters
    matrix[n,k]  X;          //n*k matrix
    vector<lower=0>[n] y; //age at death
}
```

```
parameters {
    vector[3] beta;
    real<lower=0> sigma;
}

model {
  beta[1] ~ normal(70,5);
  beta[2] ~ normal(0.01,0.001);
  beta[3] ~ normal(0,5);
  sigma ~ cauchy(0,1.25);

  y ~ normal(X*beta,sigma);
}

generated quantities{
  vector[n] y_new;
  vector[n] log_lik;
  real eta;

  for(i in 1:n){
    eta = X[i]*beta; //X[i] is ith row
    y_new[i] = normal_rng(eta,sigma);
    log_lik[i] = normal_lpdf(y[i]|eta,sigma);
  }
}
```

Vectors in Stan are, by default, column vectors. Also if 'X' is a matrix then 'X[i]' is the *i*th row of the matrix. Hence, 'X*beta' is a standard matrix × vector operation, and 'X[i]*beta' is a row vector multiplied by a column vector. If the dimensions of the vectors and matrices involved in these operations are not compatible, then Stan will produce an error. Using this notation not only is succinct but also improves the speed of the Stan execution. In later chapters, we will sometimes use this vectorized approach and sometimes not for explanatory purposes.

Summary

We started from the assertion that the National Health Service of Borgonia is a 'leveller' – that although people may have different incomes, their health should be unaffected. We have now some evidence against that assertion – since at least income does seem to be associated with age of death. First, we showed this with males only, then with males and females allowing for the possibility that there may be differences due to sex. We saw how to introduce sex as a factor and also how to introduce an interaction term between sex and income. We found that the introduction of the interaction term did not really improve the model, and that the more parsimonious model was just as good as the more complex one. Finally, we looked at using the vectorization capabilities of Stan both for elegance and for efficiency.

It is important to note that nothing in what we have done proves in any sense that income is a 'cause' of age of death. We have found an association, and one that also includes sex, but this is not causal. There are many factors involved in mortality risk including genetics, lifestyle, physical exercise, the public health system including accessibility, overall hygiene, psychological factors, the economy and interactions with personal psychology (e.g. unemployed may be more prone to depression), and pollution which may vary from region to region in the country. Also, more wealthy individuals may have access to private healthcare beyond the means of the less well off. Generally, if variables x and y are associated (correlated), this does not show any causal connection between the two – for example, there may be a third set of variables z which do have a causal effect on x and y separately, thus producing a spurious correlation between x and y. Under carefully controlled experimental conditions, there would be circumstances where it might be possible to infer causality, but generally that is not the case. Having found the relationship between age at death, income, and sex, it is up to a scientific theory to explain this.

In this chapter, we have introduced no fundamentally new concepts with respect to statistical inference in comparison to Chapter 3. We have only shown how the Bayesian framework can be easily extended to encompass the study of relationships between variables, and also we have shown how to accomplish this in the Stan programming language. The new material is concerned with model comparison, and here we have briefly discussed two methods, the Bayes factors and the methods associated with *loo*.

Online Resources

The data, R, and Stan programs that go along with this chapter are available at www.kaggle.com/melslater/slater-bayesian-statistics-4

Note that if you run the programs you may get slightly different results from what you find in this chapter. Why?

(Because it is a simulation based on random number generators so that each time you run the program, you will get a slightly different answer.)

For the 'loo' package, see https://mran.microsoft.com/snapshot/2015-10-20/web/packages/loo/vignettes/Example.html

Five

GENERAL MODELS

Introduction

In the previous chapter, we looked at some examples exploring the relationship between a response variable (age at death), an explanatory variable (income), and a factor (sex). Here we show that what we learnt can be easily generalized in various ways. There can be more than one explanatory variable. There can be more than one factor (e.g. modelling the results of an experiment). The explanatory variables and factors can be mixed together in one model. We need not rely on a normal distribution to model the explanatory variable. The relationship between the central value (e.g. mean or median) and the model involving the explanatory variables and factors need not be one of identity. There can be multiple response variables rather than a single one. A model may consist not just of one but several statistical model equations, and where the response variable in one equation could be an explanatory variable in another. We will consider each of these in turn. First, we introduce the ideas at an abstract level and then give examples.

All the examples we have considered to date have been complete, in the sense that there are no missing entries in the data. This is highly unrealistic. Real experimental and observational data typically have missing values. A simple way to deal with this is just to delete those individuals where there was missing data on important variables. Here, we present the basic idea of a more principled method for accounting for these missing data entries and a simple example.

Examples of the General Linear Model

The Typical Model

From theoretical or other considerations, it is thought that k explanatory variables $x_1, x_2, \ldots, x_k$ might influence a response variable y. A total of n independent observations are collected leading to $(y_i, x_{1i}, x_{2i}, \ldots, x_{ki})$, $i = 1, 2, \ldots, n$. These are the observations per individual i. For example, in the previous chapter, the individuals were records corresponding to individual people, where y_i was the age at death of individual i and x_{1i} was that person's income ($k = 1$). The number of individuals is clearly n. In this case, the individuals are people, but this may not be the case. For example, they could be a family unit where y_i is the average monthly expenditure on food for the ith family, x_{1i} is the number of children, and x_{2i} is the monthly net income of the household. Here, the idea might be to try to model the amount spent on food by variations in these two explanatory variables. Another example is when the individual units are a type of machine, where y_i is the time to failure from its first use, x_{1i} is a measure of the intensity of its usage, x_{2i} the average daily temperature of the environment in which it was located, and x_{3i} an overall measure of the skill of the operators who used it. This postulates a model that the time to failure is influenced by these three variables.

A typical model, but not the only possibility by far, is the following. We define the *linear predictor* of the model to be

$$\eta_i = \beta_0 + \beta_1 x_{1i} + \beta_2 x_{2i} + \ldots + \beta_k x_{ki} = \sum_{j=0}^{n} \beta_j x_{ji} \tag{5.1}$$

where $x_{0i} = 1$.

Here, the β_j are unobservable parameters, whereas the x_{ji} are known values. The term 'linear' here refers to linearity in the β_j. For example, this would also be considered as linear, since it is linear in the parameters:

$$\eta_i = \beta_0 + \frac{\beta_1}{x_{1i}} + \beta_2 x_{2i}^3$$

A way to think of this one is that the explanatory variables are actually $1/x_1$ and x_2^3.

The response variable y is typically modelled with a normal distribution:

$$y_i \sim normal(\mu_i, \sigma) \tag{5.2}$$

with mean μ_i and standard deviation σ. This may or not be a useful model, but it is typical.

Still, how do Equations (5.1) and (5.2) provide a model of how the $x_1, \ldots, x_k$ influence y, since there seems to be no connection between the response and explanatory variable?

The (typical) answer here is

$$\mu_i = \eta_i \tag{5.3}$$

The linear predictor term η_i seems to be unnecessary, since we could simply write

$$y_i \sim normal(\beta_0 + \beta_1 x_{1i} + \beta_2 x_{2i} + \ldots + \beta_k x_{ki}, \sigma)$$

This says that each y_i is a normally distributed random variable, but with mean that depends linearly on $x_1, \ldots, x_k$. However, we retain η_i since it is useful later.

It is worth mentioning here a slightly different formulation as follows:

$$y_i = \beta_0 + \beta_1 x_{1i} + \beta_2 x_{2i} + \ldots + \beta_k x_{ki} + \varepsilon_i$$

where the ε_i, $i = 1, 2, \ldots, n$, are independent normal random variables with mean 0 and standard deviation σ:

$$\varepsilon_i \sim normal(0, \sigma)$$

This is exactly the same as Equations (5.2) and (5.3) since by taking expected values, $E(y_i|\beta, \sigma) = \mu_i$. The ε_i are usually called the theoretical 'residual errors', and they are unobservable. They can be estimated by computing fitted values for the y_i from the model, and then subtracting these from the observed values y_i. For example, if we knew the posterior distributions of the β_j, we could use the means ($\hat{\beta}_j$) of each of those as estimates of the β_j and then find $\hat{y}_i = \sum_{j=0}^{k} \hat{\beta}_j x_{ji}$. The estimated residual errors would then be $e_i = y_i - \hat{y}_i$, $i = 1, 2, \ldots, n$. What is important to note from this is that the reader will sometimes come across statements that the 'dependent variable must be normally distributed'. The above shows that it is not the dependent variable (e.g. 'age at death' in the previous chapter) but the theoretical residual errors that must be normally distributed in this model. Or another way to put this is that each individual y_i conditioned on the parameters β_j and σ is normally distributed. Note that this means that unless the μ_i are all equal to one another, the response variable cannot possibly be itself normally distributed, since each of the random variables that constitute its observations has a *different* normal distribution. Note also that this model has another assumption that the standard deviation σ is the same across all individuals (which might not be the case). This is an example of an assumption called 'equal variance'.

Meaning of the Parameters

Let's consider the meaning of the β_j. Of course, these are unobservables, and even if the model were an excellent representation of reality, we could never know their 'true' values. But the point of collecting data is to find probability distributions for the β_j, which would therefore give us some idea about their possible values. The critical thing to know is whether a β_j might be inferred to be positive or negative, or another way to consider this is whether the probability distribution effectively rules out zero as a possibility. If we find that there is a high posterior probability that $\beta_j > 0$, then this implies that x_j is positively associated with y: the greater the value of x_j, the greater the value of y; for example, the more children in the family, the greater the expenditure on food. If $\beta_j < 0$, then greater values of x_j are associated with smaller values of y. For example, the greater the intensity of usage of a machine, the less the time to its failure. If, however, $\beta_j = 0$ is probable, then it means that x_j and y are not associated – for example, it may be the case that monthly income of a family does not influence its expenditure on food.

So the point of the exercise is to find posterior distributions of the β_j and then consider what the results mean for the relationships between the variables – always bearing in mind that none of this is a demonstration of *cause* between any x_j and y, as we discussed in the previous chapter.

It's not just the sign of the β_j but also the size that is important. We might find, for example, that it is highly probable that $\beta_j > 0$, but that it is so small that nevertheless variations in x_j would be associated with negligible changes in y. One way to consider this is to standardize all of the x_j and y to have mean 0 and standard deviation 1. Then β_j has a very clear interpretation: it is the number of standard deviations of change in y for a unit change in x_j, if everything else is held equal (all the other xs). Standardization is anyway useful when the x_j are measured in very different units, for example some of them in thousands and others fractions of 1.

The size of β_j is an example of an effect size, as we introduced in the previous chapter. It tells us not only whether x_j is associated with y but also how important that association might be. In the interests of parsimony, a variable with a low effect size could be eliminated from a model, except when for some reason it might be theoretically important.

Matrix Formulation

The model given in Equations (5.1) and (5.2) can be written in a very succinct form using vector and matrix notation.

Let

$$\boldsymbol{y} = \begin{pmatrix} y_1 \\ y_2 \\ y_3 \\ \vdots \\ y_n \end{pmatrix}, \boldsymbol{\beta} = \begin{pmatrix} \beta_0 \\ \beta_1 \\ \beta_2 \\ \vdots \\ \beta_k \end{pmatrix}, \text{ and } \boldsymbol{X} = \begin{pmatrix} 1 & x_{11} & x_{21} & \cdots & x_{k1} \\ 1 & x_{12} & x_{22} & \cdots & x_{k2} \\ 1 & x_{13} & x_{23} & \cdots & x_{k3} \\ \vdots & \vdots & \vdots & \vdots & \vdots \\ 1 & x_{1n} & x_{2n} & \cdots & x_{kn} \end{pmatrix}$$

Then,

$$y \sim normal(\boldsymbol{X\beta}, \Sigma) \tag{5.4}$$

X is referred to as the *design matrix*. Here Σ is the variance–covariance matrix, with $\Sigma = \sigma \mathbf{I}$, where $\mathbf{I}$ is the identity matrix. Note that this means that it is assumed that the y_i are uncorrelated (their covariances are all 0). Uniquely, in the case of the normal distribution, variables that are uncorrelated are also statistically independent as we saw in Chapter 2.

A Bayesian Model

As we have seen, the specification of a Bayesian model for inferences about the unobservable parameters (β_j and σ) requires choice of prior distributions, specification of the likelihood (probability distribution of the observations conditional on the parameters), and the posterior distributions of the parameters.

The simplest (but not the best) choice for the prior distributions is to choose improper priors: each parameter has uniform 'probability' over all possibilities $-\infty < \beta_j < \infty$, and $-\infty < \log(\sigma) < \infty$. The likelihood is given by Equations (5.1) and (5.2) – or equivalently Equation (5.4). Then, the Stan program to represent this would be as follows:

Box 5.1

```
data {
  int<lower=0> n;    //number in sample
  int<lower=0> kp1; //number of parameters plus 1
  matrix[n,kp1] X;  //n*(k+1) matrix - 1st col of X is 1 and k
                      //variables

  vector [n] y;
}

parameters {
  vector[kp1] beta;
```

(Continued)

```
  real<lower=0> sigma;
}

model {
  y ~ normal(X*beta, sigma);
}
```

Notice that the formulation of the model is just another way of writing Equation (5.4), the advantage being that this is a computer program.

In practice, as we have seen in the previous chapter, it is better to choose proper prior distributions. For example, the priors for the β_j might be normal distributions, with standard deviation dependent on the measurement scales of the variables, and for σ a distribution such as the Gamma, inverse Gamma, or Cauchy (restricted to positive values), again dependent on the scale of y.

A Normal Distribution Example

A hotly contested issue in the Borgonia education department is how reading ability of children is impacted by issues such as class size and level of socio-economic standing. Some argue that larger class sizes and economic deprivation lead to poorer reading skills amongst children, and others argue that reading ability is based on the innate intelligence of the children, and no matter what the teaching environment, and no matter what the level of socio-economic advantage or disadvantage, reading ability would not change.

Civil servants in the Department of Education carried out a study. Unfortunately, the Department of Education is underfunded to the extent that the study only involved 100 randomly selected 12-year-old children from across the country. The response variable measured was the reading age of the child, which is a number that reflects reading ability based on independent assessment of the child and a battery of tests. For example, a 12-year-old with the reading ability score of 12 is average, with a score of less than 12 is behind, and with a score of greater than 12 is advanced. Past data indicate that scores are approximately normally distributed with mean 12 and standard deviation of about 4.

The explanatory variables measured were the average class size of the child's school, annual gross income of the family, and the historically recorded IQs of both parents (available only to the civil servants).

Figure 5.1 shows the histograms of the response and explanatory variables. Reading age shows some possible outliers at the higher end. The means and standard deviations are shown in Table 5.1. The mean reading age is slightly above the age 12 average, but with quite large variability.

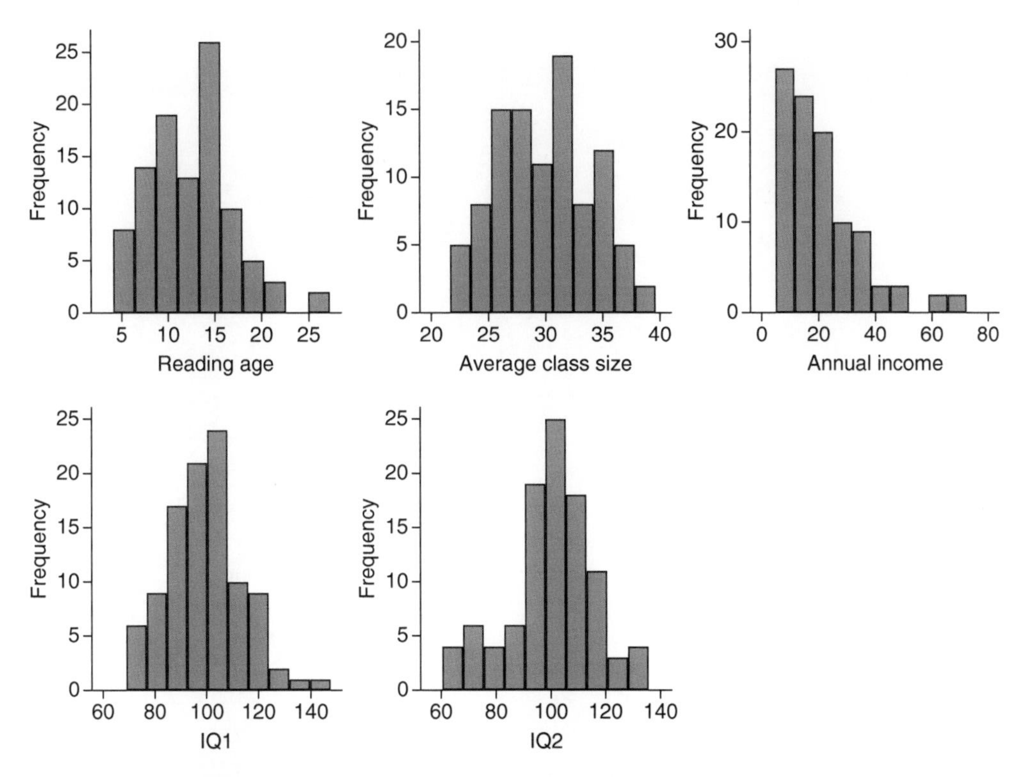

Figure 5.1 Histograms of the response variable reading age and the explanatory variables. Annual income is in 1000s

Table 5.1 Means and standard deviations of the variables

	Reading age	Class size	Income (in 1000s)	IQ1	IQ2
Mean	12.6	30.0	21.4	99.1	100.2
SD	4.50	4.08	14.13	14.47	15.53

Note: IQ = intelligence quotient.

Figure 5.2 shows the scatter diagrams of reading age on the explanatory variables. There appears to be a negative correlation with class size, a stronger positive correlation with income, and possible, but weak, correlations with the IQs of the parents (at least for IQ1). For prior distributions for the β_j, a normal distribution with mean 0 and standard deviation 10 is used. This gives a wide range of possibilities, since the 95% prior credible interval is –20 to 20. For σ, a Gamma distribution with parameters 2 and 0.1 is used, which has 95% prior credible interval approximately 2.4 to 55.7. This is shown in Figure 5.3.

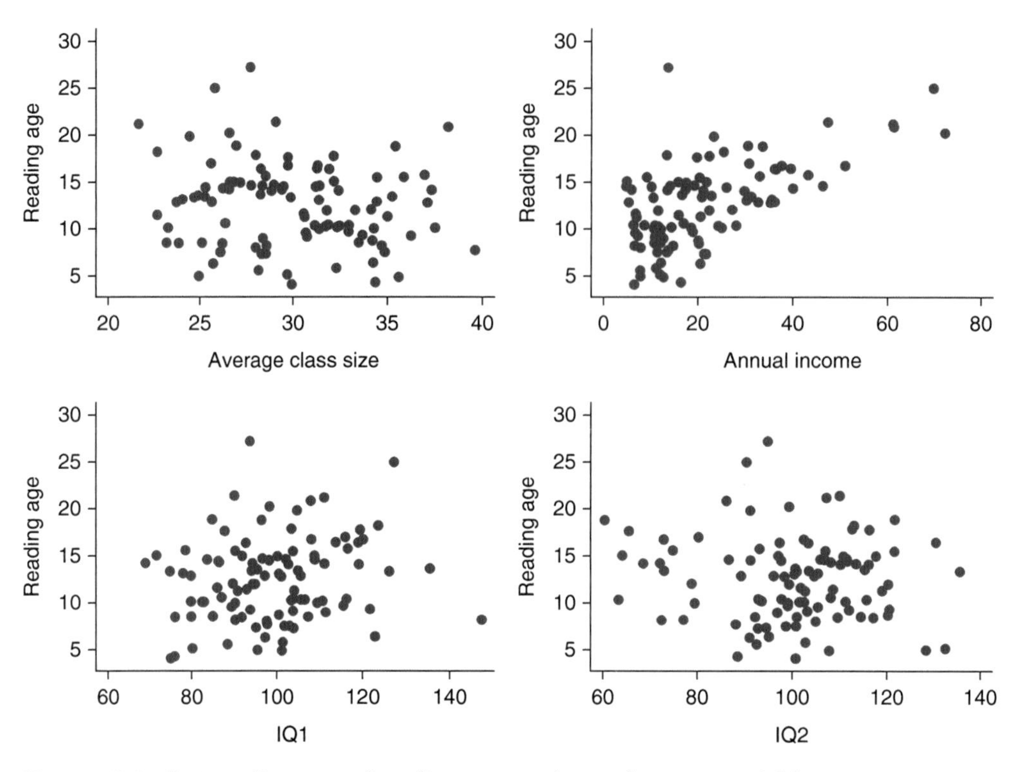

Figure 5.2 Scatter diagrams of reading age on the explanatory variables

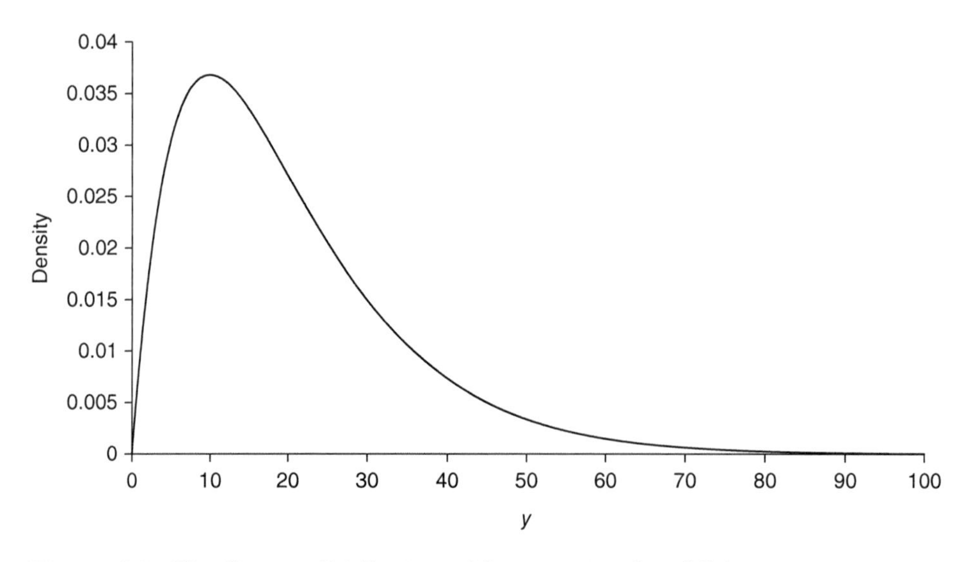

Figure 5.3 The Gamma distribution with parameters 2 and 0.1

Table 5.2 shows a summary of the posterior distribution of the parameters, including the posterior 95% credible interval. Recalling that for the β_j the prior intervals were ±20, we can see a very considerable narrowing. Similarly for σ. We can see that it is highly probable that $\beta_0 > 0$. The probability $\beta_1 < 0$ is $1 - 0.081 = 0.919$. From the mean of the distribution of β_1, we can see that on the average an increase in class size of 1 is associated with a reduction of reading age by 0.13. That doesn't sound too important. But it also means that an increase in class size by 10 is associated with a reduction in reading age by just over 1 year, which is important. So class size, while not having an extremely strong probabilistic association with reading age, has a strong effect size. Income is almost certain to have a positive association with reading age. An increase of income by 10,000 is associated with an increase in reading age by almost 2 years. IQ1 is likely positively associated with reading age (the probability of its being positive is just under 0.9). However, its effect size is very low. An increase in IQ by 10, for example, is associated with an increase in reading age only by 0.3. IQ2 does not appear to be associated with reading age at all. Unfortunately, the civil servants will not release the gender of IQ1 or IQ2, as they say 'for reasons of avoiding gender-biased inferences' about the relative roles of the parents.

Table 5.2 Summaries of posterior distributions of the parameters for the normally distributed likelihood model

Parameter	Coefficient of	*M*	*SD*	2.5%	97.5%	Prob > 0
β_0		10.86	3.92	3.07	18.57	0.997
β_1	Class size	−0.13	0.09	−0.30	0.05	0.081
β_2	Income	0.18	0.03	0.13	0.24	1.000
β_3	IQ1	0.03	0.03	−0.02	0.09	0.895
β_4	IQ2	−0.02	0.02	−0.06	0.03	0.234
σ		3.65	0.27	3.17	4.23	

Note: Prob > 0 is the posterior probability that the parameter is > 0. IQ = intelligence quotient.

Recall that in the 'generated quantities' segment of the Stan program, we can generate the posterior predicted values for $y_1, y_2, \ldots, y_n$. Figure 5.4 shows the means of these distributions plotted against the observed values. Although the plot shows a strong relationship between the observed values and posterior predicted means ($r = 0.63$), there are some problems. The most notable is the outlier at the highest value of the observed reading age, and also the distribution shows some skewness with respect to a hypothetically fitted straight line. We can see possible outliers also in Figures 5.1 and 5.2, and this result probably reflects those.

Getting Away From the Normal

In classical statistics, assumptions of normality are the norm, and this is mainly because the mathematical theory and the computational methods are most tractable with this assumption. The theory of multiple regression assumes, as we saw in the last section, that each

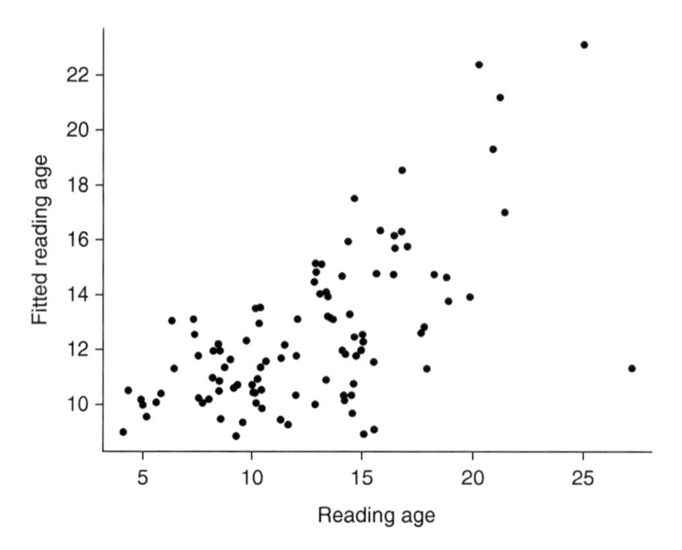

Figure 5.4 Means of the distributions of the fitted values by observed values of reading age

individual observation on the response variable is normally distributed with mean equal to the linear predictor. In classical statistics, if evidence emerges that the normality assumption is not supported, then various ad hoc techniques are used – such as transforming variables (e.g. taking logs or square roots) – in order to try to find a representation of the data where the normal assumption is acceptable. Of course, there are circumstances where another distribution is indicated on theoretical grounds (as we will see in the later sections). But for standard 'multiple regression', the normal assumption is essentially unavoidable if the classical results are required, such as t tests on the parameters.

However, in the Bayesian method, we can easily select other distributions. We suppose that the type of distribution that normality represents (symmetric and approximately bell-shaped around the mean) is appropriate. But then, we would not expect the outliers observed. What we need is another distribution, like the normal, but which is more 'spread out' – where high probabilities might be assigned to values far from the central one, while maintaining the type of shape that the normal distribution has. Recalling Chapter 2, there is the Student t distribution, which has an additional parameter as the degrees of freedom (ν). For large values of ν (around 30 and more), the Student t distribution approximates the normal distribution closely. So here we have the best of all worlds – we get away from strict reliance on normality, preserve the symmetry of the distribution, but also allow for the possibility that the appropriate distribution might even be normal if ν turns out to be large.

If we select the likelihood to be Student's t, how do we choose the value of ν? This is simple: ν is just another unobservable parameter. We give it a prior distribution, and its posterior distribution is found using the data.

The Stan program is then, with changes compared to the normal case highlighted in red,

Box 5.2

```
data {
  int<lower=0> n;         //number in sample
  int<lower=0> k;         //number of parameters
  matrix[n,k]  X;         //n*k matrix
  vector<lower=0>[n] y; //reading age
}

parameters {
  vector[k] beta;
  real<lower=0> sigma;
  real<lower=1> df;
}

model {
  for(i in 1:k)
    beta[i] ~ normal(0,10);

    sigma ~ gamma(2,0.1);
    df ~ gamma(2,0.1);

    y ~ student_t(df, X*beta, sigma);
}

generated quantities{
  vector[n] y_new;
  vector[n] log_lik;
  real eta;

  for(i in 1:n){
    eta = X[i]*beta; //X[i] is ith row
    y_new[i] = student_t_rng(df, eta, sigma);
    log_lik[i] = student_t_lpdf(y[i]|df, eta, sigma);
  }
}
```

Note that the prior for the degrees of freedom is also the Gamma distribution (2,0.1). This gives an approximately 20% prior probability that $df > 30$.

Table 5.3 shows the summary of the posterior distributions. It is almost the same as Table 5.2. Note that the mean of the posterior distribution of the *df* is just under 17, which is well below what would be expected for the normal distribution. The posterior distribution for *df* is shown in Figure 5.5. The posterior probability that $df > 30$ is 0.11.

Table 5.3 Summaries of posterior distributions of the parameters for the Student *t* distributed likelihood model

Parameter	Coefficient of	*M*	*SD*	2.5%	97.5%	Prob > 0
β_0		10.32	3.70	2.94	17.70	0.996
β_1	Class size	−0.12	0.08	−0.28	0.04	0.073
β_2	Income	0.19	0.02	0.14	0.24	1.000
β_3	IQ1	0.03	0.02	−0.01	0.08	0.912
β_4	IQ2	−0.02	0.02	−0.06	0.03	0.247
σ		3.25	0.30	2.69	3.88	
df		16.88	10.99	4.77	46.50	

Note: Prob > 0 is the posterior probability that the parameter is > 0. IQ = intelligence quotient; *df* = degrees of freedom.

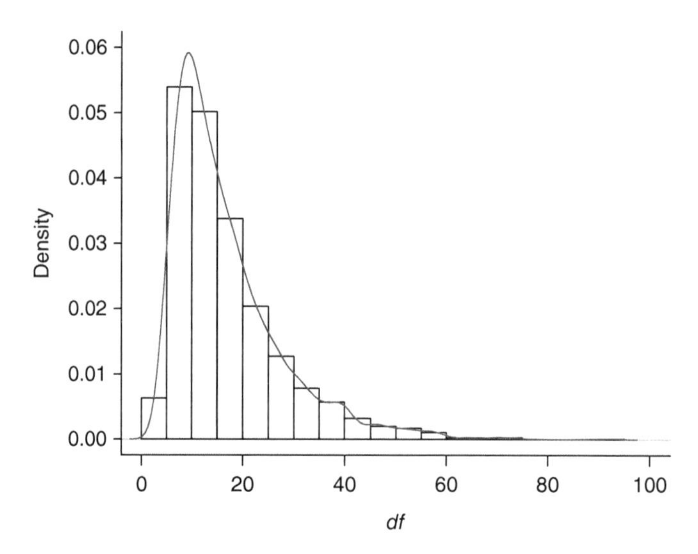

Figure 5.5 The posterior distribution of degrees of freedom (*df*) shown as a histogram and the fitted density function

The equivalent plot for Figure 5.4 looks almost the same so that it seems that using the Student *t* model has no advantage.

Let's now compare the two models using the 'leave-one-out' (loo) method.

Table 5.4 shows the estimates, and it shows that the Student *t* has the greater expected log pointwise predictive density (*elpd*) but that there is no substantial difference between the two.

Table 5.4 Comparing (A) the two models individually and (B) the difference between them

(A) Each individual model

	Normal		Student t	
	Estimate	***SE***	**Estimate**	***SE***
elpd_loo	–273.4	11.4	271.3	9.5
p_loo	6.2	2.1	6.5	1.4

(B) Comparison between the models

	elpd_diff	**se_diff**
Student t	0.0	0.0
Normal	–2.1	2.8

Note: elpd = expected log pointwise predictive density; loo = leave-one-out; diff = difference; se = standard error.

However, the loo function indicates that one individual in the normal model has Pareto k value too high. This is shown in Figure 5.6. It can be seen that in the case of the normal model, individual 80 has a high k value, whereas for the Student t model, the Pareto k values are smaller overall.

Although in terms of the elpd there is not much to choose between the two models, overall the fit for the Student t distribution is preferable.

The posterior predicted value for individual 80 is far too high for either model. For example, in the case of the Student t model, the probability that this value would be greater than 20 is 0.009, and this is even smaller for the normal distribution model. Hence, the data value for individual 80 is suspect, and in an actual situation should be carefully checked for error (which could have occurred at many different times, from the original data collection, through to transcribing the data from one source to another).

Here, we relaxed the assumption that the likelihood function has to be based on a normal distribution. Analogously, in frequentist statistics, there are methods that allow relaxation of the classical assumptions of normality and equal variance, generally referred to as 'robust regression' techniques – robust because they are less sensitive, for example, to outliers. For a very comprehensive account of classical regression analysis, see Montgomery et al. (2021), where robust regression is covered in Chapter 15.

In all of the above, still the link between the mean of the distribution and the linear predictor is the identity. In the next section, we examine some models where this is no longer the case.

Generalized Linear Models

The Formulation

The model stated in Equations (5.1) and (5.2) is called the 'general linear model'. The response variable has a mean which is equal to the linear predictor, and the observations are independent

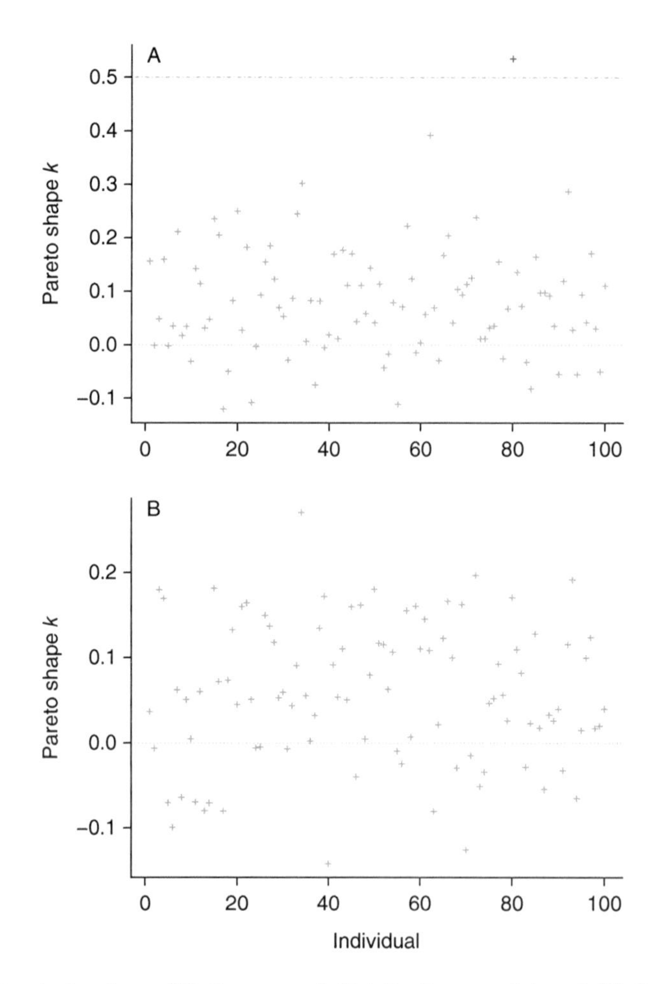

Figure 5.6 Pareto k values: (A) the normal distribution model and (B) the Student t model

samples from a normal distribution, and the standard deviation is constant. Equation (5.4) provides a very succinct way to write this model. It is called the 'general' linear model because, depending on how the values in the $\boldsymbol{X}$ (design) matrix are specified, this one model accounts for many different types of analysis: multiple regression, analysis of variance, and analysis of covariance.

In this section, we generalize beyond the general linear model to the *generalized linear model* (McCullagh and Nelder, 1989). This has three components:

1. A random variable y with independent observations $(y_1, y_2, ..., y_n)$
2. The expected value and variance, which are $E(y_i) = \mu_i$ and $Var(y_i) = \sigma^2$

3. The linear predictor $\eta_i = \sum_j \beta_j x_{ji}$

So far so good with respect to the general linear model. However,

4. The link function $g(\mu_i) = \eta_i$

In the general linear model, the *link function* is the identity (Equation 5.3). In the generalized linear model, the identity is only a special case, it can equally well be another function, such as log (for example).

The generalized linear model is therefore more 'general' than the general linear model, certainly with respect to the link function. However, there is another level of generality, which is that the normal distribution is also only a special case. The response variable y is typically a distribution that belongs to the exponential family of distributions:

$$f(y|\theta,\phi) = \exp\left(\frac{y\theta - b(\theta)}{a(\phi)} + c(y,\phi)\right) \quad (5.5)$$

Here a, b, and c are specific functions and (θ,ϕ) are parameters. It is not difficult to show that

$$E(y|\theta,\phi) = b'(\theta)$$

and

$$var(y|\theta,\phi) = a(\phi)b''(\theta) \quad (5.6)$$

where b' is the derivative with respect to θ and b'' is the second derivative.

When the link function can be expressed in this formulation,

$$\theta = \eta \quad (5.7)$$

then this is called the *canonical link*. More specifically, the canonical link is the link function that satisfies Equation (5.7). In reality, any link function that makes sense in terms of the underlying data can be used, but when specifically it satisfies Equation (5.7), it is the canonical link. In classical statistical theory, choice of the link as the canonical link leads to simpler formulations and well-understood desirable properties for estimation of the unobservable parameters.

The Normal Distribution

The exponential family in Equation (5.5) looks quite different from the normal distribution:

$$f(y|\mu,\sigma^2) = \frac{1}{\sigma\sqrt{2\pi}}\exp\left(-\frac{1}{2}\left(\frac{y-\mu}{\sigma}\right)^2\right)$$

However, if we square out the bracket in the exponential, we can rearrange this as

$$f(y|\mu,\sigma^2) = \exp\left\{\left(\frac{y\mu - \frac{\mu^2}{2}}{\sigma^2}\right) + c(y,\sigma^2)\right\}$$

Hence,

$$\theta \equiv \mu$$

$$\phi \equiv \sigma^2$$

and $a(\phi) = \phi$.

$$b(\mu) = \frac{\mu^2}{2}$$

$$b'(\mu) = \mu$$

$$b''(\mu) = 1$$

Hence,

$$E(y|\theta,\phi) = b'(\theta) = \mu$$

$$var(y|\theta,\phi) = a(\phi)b''(\theta) = \sigma^2$$

Moreover the identity link function $\mu = \eta$ is also the canonical link since this is the same as Equation (5.7).

The Poisson Distribution

Recall from Chapter 2 the Poisson distribution

$$f(y|\mu) = \frac{\mu^y}{y!}\exp(-\mu)$$

Again this looks very different from the exponential family shown in Equation (5.5). However, bringing everything under the exponential,

$$f(y|\mu) = \exp(-\mu + y\log(\mu) - \log(y!))$$

Setting $\theta = \log(\mu)$, then $\mu = \exp(\theta)$ so that $b(\theta) = \exp(\theta)$ results in Equation (5.5), with $a(\phi) = 1$.

Hence,

$$E(y|\mu) = b'(\theta) = e^{\theta} = \mu$$

$$var(y|\mu) = b''(\theta) = \mu$$

following what we know from the Poisson distribution that the mean and variance are the same. For the canonical link function, from Equation (5.7)

$$\theta = \log(\mu) = \eta$$

The formulation $\log(\mu) = \eta$ makes sense from another point of view, not just the desired statistical properties that accompany a canonical link. Of course, this is equivalent to $\mu = e^{\eta}$. If we choose the identity link $\mu = \eta$, then we would have the problem that the estimators for the β_j might be such that $\eta < 0$, and hence, $\mu < 0$. But the Poisson distribution is for non-negative observations $y_i \geq 0$, so that a negative mean is impossible.

The Binomial Distribution

Let's also consider the binomial distribution

$$f(y|p) = \binom{n}{y} p^{y} (1-p)^{n-y}$$

This also looks nothing like the formulation in Equation (5.5). However, since $p^y = \exp(y\log(p))$,

$$f(y|p) \propto \exp(y\log(p) + (n-y)\log(1-p))$$

and rearranging, we get

$$f(y|p) \propto \exp\left(y\log\left(\frac{p}{1-p}\right) - n\log\left(\frac{1}{1-p}\right)\right)$$

Comparing with Equation (5.5),

$$\theta = \log\left(\frac{p}{1-p}\right)$$

and therefore,

$$p = \frac{e^{\theta}}{1+e^{\theta}}$$

$$b(\theta) = n \log(1 + e^{\theta})$$

$$b'(\theta) = n\frac{e^{\theta}}{1+e^{\theta}} = np$$

$$b''(\theta) = n\frac{e^{\theta}}{\left(1+e^{\theta}\right)^2} = np(1-p)$$

This, of course, reproduces the results we found in Chapter 2. The canonical link function is therefore

$$\log\left(\frac{p}{1-p}\right) = \eta$$

This expresses the linear predictor as the *logit* function, the log odds of 'Success' (probability p) and 'Failure' (probability $1 - p$). Further insight can be gained by considering the inverse of the link function:

$$p = \frac{e^{\eta}}{1+e^{\eta}} = \frac{1}{1+e^{-\eta}}$$

Note that this is a sensible choice of link function since whatever the value of η (as determined by the estimates of β_j), it is always the case that p is constrained to be in the interval [0, 1] (as required since p is a probability). The expression on the right-hand side is called the *logistic function.*

Note that although from the point of view of classical statistics the canonical link function is desirable, it is not necessary. For example, a different link function used in the case where the likelihood is the binomial distribution is the *probit* link. This is simply expressed in inverse form:

$$p = \o(\eta)$$

where $\o$ is the distribution function of the standard normal distribution, that is

$$\o(\eta) = P(Z < \eta)$$

where $Z \sim$ *normal*(0, 1). Whatever the value of η, the probability p will always be in the range [0, 1].

A Poisson Model for Bystander Responses to Soccer Violence

The issue of bystander responses to violent incidents has a long history in social psychology. Under what conditions do people intervene, or not, when they are witness to a violent attack of one person or group on another? There are various theories about 'bystander' behaviour; in particular, the most widespread and accepted theory goes back to a true case in New York in 1964, where according to the story, many people watched the murder of a young girl and did nothing to help.[1] The theory, based on numerous observations, is that the more people watching a violent event, the less likely it is that anyone of those bystanders will intervene to help (Darley and Latané, 1968; Latané and Darley, 1968). Others have argued that it is a matter of 'social identity' – the more that bystanders identify with the victim, the more likely they are to intervene (Levine and Manning, 2013).

Depicting a violent confrontation in real life for experimental studies is ethically impossible, so researchers have used virtual reality. In a study that we carried out, we recruited 40 participants who were all strong fans of the UK Arsenal soccer club (Slater et al., 2013). The setting was a bar. In the virtual reality, first of all a virtual human character (V) talked to the participant about soccer. After a while, another virtual human character (P) entered and began to verbally and eventually physically attack V. The human participant was the bystander. The question was whether during this confrontation, the participant would attempt to help V or not.

There were two conditions in the experiment: Group and Gaze. The InGroup condition (Group = 1) was when V was also an enthusiastic Arsenal supporter, and the OutGroup condition (Group = 0) was when V was just a neutral soccer fan, but not an Arsenal fan. The Gaze condition involved V occasionally looking towards the participant during the confrontation (LookAt, Gaze = 1) and where V did not look towards the participant (NoLookAt, Gaze = 0).

Table 5.5 shows how the participants in the study were (randomly) assigned to the conditions. Unfortunately, as happens, data were lost on two participants, so finally there were only 38. We will return to missing values later.

Table 5.5 Numbers of participants in the conditions of the bystander study

Group	**NoLookAt**	**LookAt**	**Total**
OutGroup	9	9	18
InGroup	10	10	20
Total	19	19	38

The behaviour of the participants was video recorded, and independent experts assessed whether and how many interventions participants carried out to help V in any way during the confrontation. They did this assessment 'blindly' without knowing the experimental condition. In particular, they recorded two variables: *numphysical*: y_{phys}, which was the number of times that each participant intervened physically during the argument, such as stepping between V and P. Similarly, *numverbal*: y_{verb} was the number of verbal interventions to help V made by the participants such as exclaiming 'Leave him alone!' and so on. The interest is on whether the experimental factors Group and Gaze influenced the intervention responses

[1]https://en.wikipedia.org/wiki/Murder_of_Kitty_Genovese

of the participants. In addition, two other response variables were recorded: *timephysical*: t_{phys} and *timeverbal*: t_{verb}; these are the times in seconds in which participants engaged in physical or verbal interventions.

Figure 5.7 shows the means and standard errors of the intervention variables. This suggests that the Group variable influences the number of interventions – they are higher for the InGroup condition than the OutGroup. However, the Gaze condition seems to have no effect.

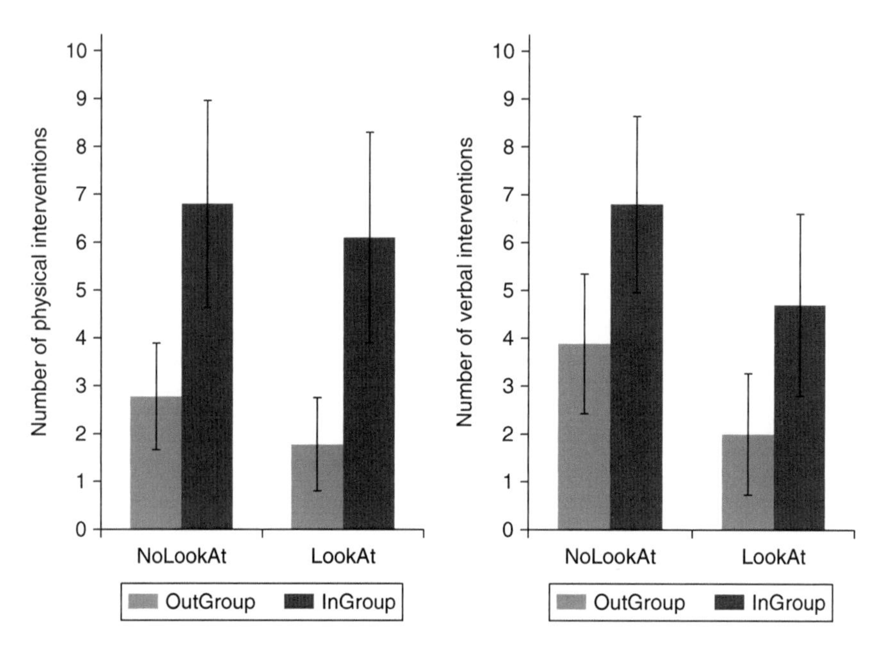

Figure 5.7 Means and standard errors of the number of physical and verbal interventions

First, we concentrate solely on the number of physical interventions. In order to analyse these data, we need to choose the distribution of the numbers of interventions for the likelihood. Figure 5.8 shows the histograms of the numbers of physical and verbal interventions by Group. Remember that these are count variables corresponding to events that occasionally occur through time. The first thought about a distribution to model this situation is the Poisson distribution. In this case,

$$y_{phys,i} \sim Poisson(\mu_i),\ i = 1, 2, \ldots, n = 38$$

and we know from the discussion above that an appropriate link function would be

$$\log(\mu_i) = \eta_i$$

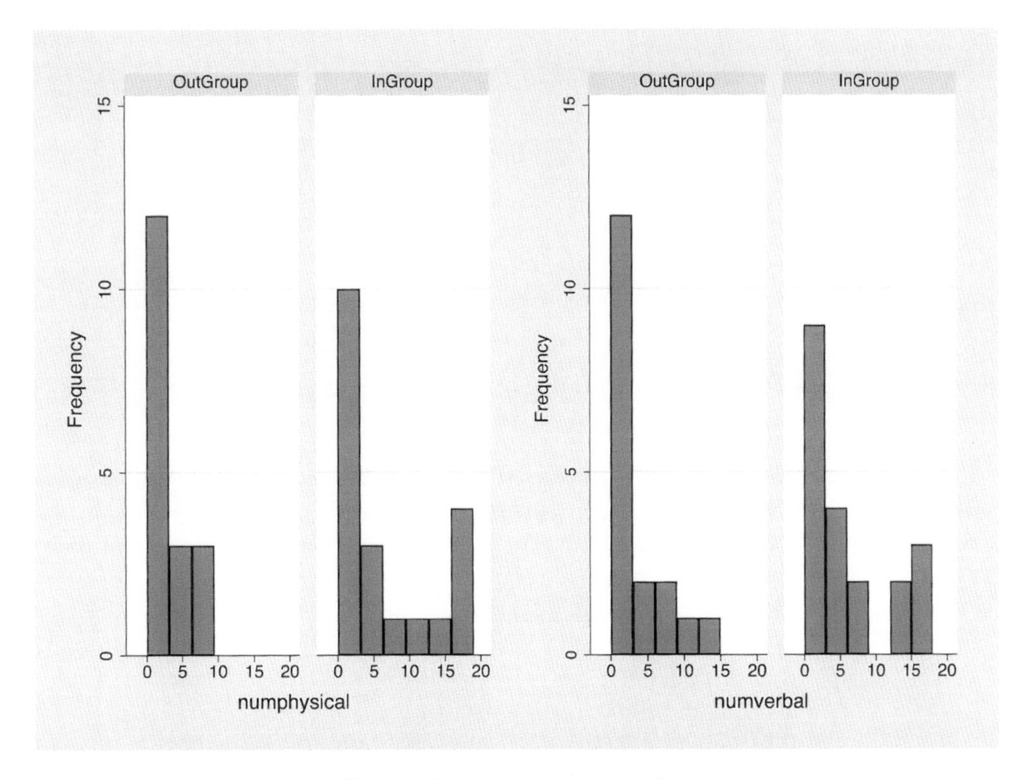

Figure 5.8 Histograms of the number of interventions by Group

The linear predictor, taking account of the two factors and their interaction, is

$$\eta_i = \beta_{phys,0} + \beta_{phys,1} g_i + \beta_{phys,2} z_i + \beta_{phys,3}(g_i z_i) \tag{5.8}$$

where

$$g_i = \begin{cases} 0, & \text{OutGroup} \\ 1, & \text{InGroup} \end{cases}$$

$$z_i = \begin{cases} 0, & \text{NoLookAt} \\ 1, & \text{LookAt} \end{cases}$$

Let's consider an implication of this model. From the link function,

$$\log(\mu_i) = \beta_{phys,0} + \beta_{phys,1} g_i + \beta_{phys,2} z_i + \beta_{phys,3}(g_i z_i)$$

Exponentiating each side of this equation (to get the 'inverse link'),

$$\mu_i = (e^{\beta_{phys,0}})(e^{\beta_{phys,1}g_i})(e^{\beta_{phys,2}z_i})(e^{\beta_{phys,3}(g_i z_i)})$$

Now compare what happens when $g_i = 0$ or $g_i = 1$ in the case that $z_i = 0$:

$$\mu_i(g_i = 0, z_i = 0) = (e^{\beta_{phys,0}})$$

$$\mu_i(g_i = 1, z_i = 0) = (e^{\beta_{phys,0}})(e^{\beta_{phys,1}})$$

Now consider the ratio

$$\frac{\mu_i(g_i = 1, z_i = 0)}{\mu_i(g_i = 0, z_i = 0)} = e^{\beta_{phys,1}}$$

If $\beta_{phys,1} < 0$, then the mean at $g_i = 1$ is less than when $g_i = 0$. If $\beta_{phys,1} = 0$ then the two means are equal, and if $\beta_{phys,1} > 0$, then the mean at $g_i = 1$ is greater than when $g_i = 0$. The reader is advised to repeat this analysis for the case $z_i = 1$. Care must be taken in interpreting the parameters in the case of non-identity link functions.

Another typical notation used to write this model is

Group + Gaze + Group × Gaze

Group and Gaze are the two main effects and Group × Gaze is the interaction effect.

As before, the prior distributions are chosen as $\beta_j \sim$ *normal*(0, 10). When we fit this model, we obtain the results shown in Table 5.6. The results show that Group is important: it is positively associated with the number of interventions in the case when Group = 1 (InGroup). Gaze (LookAt) has probability 1 – 0.083 = 0.917 of being negatively associated with the number of interventions. However, there is less evidence for an interaction effect. Here, the results suggest that in the situation when Group = InGroup and Gaze = LookAt, there may be a positive association with the number of interventions.

Table 5.6 Summaries of posterior distributions of the parameters for Poisson distributed likelihood model

Parameter	Coefficient of	*M*	*SD*	2.5%	97.5%	Prob > 0
β_{phys_0}		1.00	0.20	0.58	1.38	1.000
β_{phys_1}	Group	0.91	0.24	0.45	1.38	1.000
β_{phys_2}	Gaze	–0.45	0.33	–1.11	0.18	0.083
β_{phys_3}	Group × Gaze	0.34	0.37	–0.36	1.10	0.820

Note: Prob > 0 is the posterior probability that the parameter is > 0.

However, the fit of the model is not good. If we carry out the 'loo' analysis, we find, for example, that the effective number of parameters (p_loo) is 21, whereas there are only four actual parameters, and also there is one Pareto k value that is classified as 'bad'. This corresponds to one very high number of physical interventions compared to the remainder. In fact, we know that the Poisson distribution should have mean and variance equal, but in this case the mean and standard deviations of the numbers of interventions are 2.2 ± 3.06 in the OutGroup condition and 6.4 ± 6.67 in the InGroup. Clearly, the means and variances are nowhere near each other. This situation is referred to as *overdispersion* in the Poisson model.

To overcome this, we can move to another distribution. The negative binomial distribution (Chapter 2) can be parameterized with two parameters μ and ϕ, where, if $y \sim negative\_binomial(\mu,\phi)$, then

$$E(y \mid \mu, \phi) = \mu$$

$$Var(\mu, \phi) = \mu + \frac{\mu^2}{\phi}$$

If ϕ is large compared to μ^2, then this distribution approaches the Poisson. However, small values of ϕ allow for departure from the Poisson with greater variance. If we use this model instead of the Poisson, then we obtain the summary of posterior distributions as shown in Table 5.7. The results reaffirm that Group (InGroup) is positively associated with the number of interventions (probability 0.938), but it does not support any influence of the Gaze factor or the interaction.

Table 5.7 Summaries of posterior distributions of the parameters for the negative binomially distributed likelihood model

Parameter	Coefficient of	*M*	*SD*	2.5%	97.5%	Prob > 0
β_{phys_0}		1.08	0.46	0.26	2.04	0.995
β_{phys_1}	Group	0.91	0.61	−0.28	2.13	0.938
β_{phys_2}	Gaze	−0.45	0.67	−1.77	0.90	0.230
β_{phys_3}	Group × Gaze	0.34	0.89	−1.45	2.04	0.659
ϕ		0.76	0.24	0.38	1.33	

Note: Prob > 0 is the posterior probability that the parameter is > 0.

Figure 5.9 shows the posterior distribution of ϕ where it can be seen that the bulk of the probability is concentrated on values less than 1.

The 'loo' analysis also produces much better results. The effective number of parameters is only 4.6, which corresponds to the actual five parameters. Table 5.8 shows the result of the loo analysis, and clearly the negative binomial is the preferred model. Moreover, for the Poisson model, there are 34 'good' Pareto k estimates, 3 'ok', and 1 'bad'. But for the negative binomial, there are 37 'good' estimates and 1 'ok'.

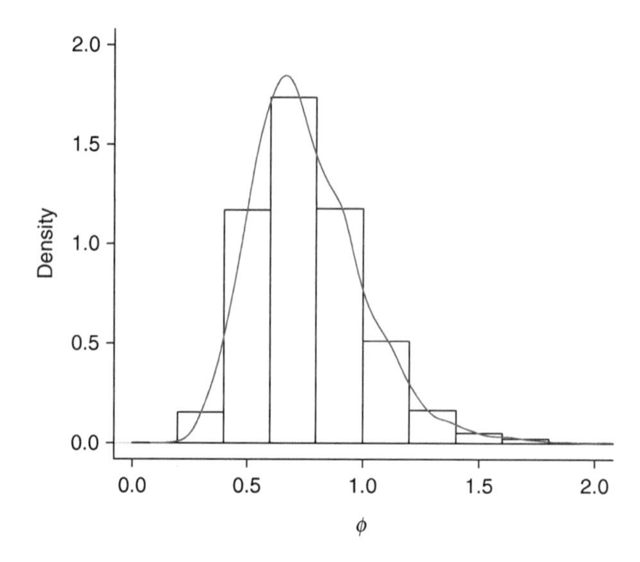

Figure 5.9 The posterior distribution of ϕ

Table 5.8 Comparing (A) the two models individually and (B) the difference between them

(A) Each individual model

	Poisson		Negative binomial	
	Estimate	***SE***	**Estimate**	***SE***
elpd_loo	–160.0	17.3	–99.3	7.8
p_loo	21.4	4.2	4.6	0.9

(B) Comparison between the models

	elpd_diff	**se_diff**
Negative binomial	0	0
Poisson	–60.7	13.0

Note: elpd = expected log pointwise predictive density; loo = leave-one-out; diff = difference; se = standard error.

Multiple Response Variables

How about the number of verbal interventions? Do we have to carry out a separate analysis for this response variable? In fact, we only have to include it in the same overall model. In other words, simultaneously in one model we can include both:

Box 5.3

```
data {
  int<lower=0> n;       //number in sample
  int<lower=0> k;       //number of parameters
  vector<lower=0,upper=1>[n] g;  //Group condition
  vector<lower=0,upper=1>[n] z;  //gaze condition
  int<lower=0> yphys[n];         //num physical
  int<lower=0> yverb[n];         //num verbal
  vector<lower=0>[n] tphys;      //time physical
  vector<lower=0>[n] tverb;      //time verbal
}

parameters {
  vector[k] b_phys;
  vector[k] b_verb;
  real<lower=0> phi1;
  real<lower=0> phi2;
}

model {
  real mu1;
  real mu2;

  for(i in 1:k) {
    b_phys[i] ~ normal(0,10);
    b_verb[i] ~ normal(0,10);
  }
  phi1 ~ gamma(2,0.1);
  phi2 ~ gamma(2,0.1);

  for(i in 1:n){
        mu1 = exp(b_phys[1] + b_phys[2]*g[i] + b_phys[3]*z[i] +
            b_phys[4]*g[i]*z[i]);
        mu2 = exp(b_verb[1] + b_verb[2]*g[i] + b_verb[3]*z[i] +
            b_verb[4]*g[i]*z[i]);
        yphys[i] ~ neg_binomial_2(mu1,phi1);
        yverb[i] ~ neg_binomial_2(mu2,phi2);
  }
}

generated quantities{
  int yphys_new[n];
```

(Continued)

```
  int yverb_new[n];
  vector[n] log_lik_phys;
  vector[n] log_lik_verb;
  real mu1;
  real mu2;

  for(i in 1:n){
    mu1 = exp(b_phys[1] + b_phys[2]*g[i] + b_phys[3]*z[i] +
        b_phys[4]*g[i]*z[i]);
    mu2 = exp(b_verb[1] + b_verb[2]*g[i] + b_verb[3]*z[i] +
        b_verb[4]*g[i]*z[i]);
    yphys_new[i] = neg_binomial_2_rng(mu1,phi1);
    yverb_new[i] = neg_binomial_2_rng(mu2,phi2);
    log_lik_phys[i] = neg_binomial_2_lpmf(yphys[i]|mu1,phi1);
    log_lik_verb[i] = neg_binomial_2_lpmf(yphys[i]|mu2,phi2);
  }
}
```

In classical statistics, when we have multiple response variables each with its own parameter set, we quickly get into trouble. Typically, separate analyses must be carried out for each model. But then, this raises the issue of 'significance'. A test might be set up with significance level 0.05. But then, what is the overall significance if there are 2, 3, 4, ..., 10 tests each with significance 0.05? It is certainly no longer 0.05. All sorts of ad hoc methods have been proposed and used to overcome this problem. In Bayesian statistics, this problem does not occur. The result of any Bayesian analysis is the joint distribution of all the parameters. We can extract as many probability statements that we like from this joint distribution without losing anything. It is as if there is a page of text, and in the classical approach (by analogy), the more information we gather from that page of text, the less reliable it is. But in the Bayesian method, of course, we can read information from that page without compromising anything. This is a fundamental advantage of the Bayesian method in the situation of multiple response variables.

As can be seen from Table 5.9, the results for the physical interactions are unchanged. For the verbal interactions, the results for the Gaze and interaction terms show relatively low probabilities of a relationship, whereas there is a moderate probability (0.817) that Group (InGroup) is positively associated with the number of verbal interactions. The loo analysis shows good results, with effective number of parameters approximately the same as the actual, and the Pareto k parameters almost all 'good' and a total of 3 'ok' values.

In practice, the number of physical and verbal interventions is correlated, and this is not taken account of in the model. This could be done by specifying a joint probability distribution over both variables for the likelihood or including one variable as an explanatory variable for the other. For example, it could be argued that a verbal interaction comes first, and a physical

Table 5.9 Summaries of posterior distributions of the parameters for the negative binomially distributed likelihood model for both numbers of physical and verbal interactions

Parameter	Coefficient of	*M*	*SD*	2.5%	97.5%	Prob > 0
β_{phys_0}		1.07	0.45	0.25	2.04	0.995
β_{phys_1}	Group	0.91	0.62	−0.33	2.15	0.934
β_{phys_2}	Gaze	−0.45	0.65	−1.70	0.84	0.235
β_{phys_3}	Group × Gaze	0.34	0.87	−1.41	1.98	0.659
ϕ_{phys}		0.77	0.24	0.40	1.31	
β_{verb_0}		1.43	0.47	0.58	2.42	1.000
β_{verb_1}	Group	0.56	0.64	−0.73	1.82	0.817
β_{verb_2}	Gaze	−0.66	0.69	−2.04	0.71	0.163
β_{verb_3}	Group × Gaze	0.28	0.93	−1.56	2.16	0.618
ϕ_{verb}		0.69	0.23	0.35	1.22	

Note: Prob > 0 is the posterior probability that the parameter is > 0.

interaction simply accompanies it, so that the physical interactions might have the verbal interactions as another explanatory variable. Moreover, we could also include the times for physical and verbal as additional response variables, thus in one overall model accounting for all of the response variables. We leave this as an exercise for the reader.

An Alternative Parameterization

Table 5.5 sets out the experimental design consisting of two factors (Group and Gaze) each at two levels. The design is 'between-groups' meaning that each one of the four cells of the experimental design is based on a different sample of participants, so that each participant had only one experience. A 'within-groups' design is where each participant may experience a factor at each of several levels (to be discussed in Chapter 7).

In classical statistics, a two-factor, between-groups, experimental design would be treated by an 'analysis of variance' (ANOVA) model. This is as follows, in terms of the linear predictor:

$$\eta_{ij} = \mu + \alpha_i + \beta_j + \gamma_{ij} \tag{5.9}$$

where i refers to the ith row of the design (Group: OutGroup = 1, InGroup = 2), and j refers to the jth column (Gaze: NoLookAt = 1, LookAt = 2) of Table 5.5. Then μ refers to a 'general mean', α_i is the effect of being in the ith row, β_j is the effect of being in the jth column, and γ_{ij} is an interaction effect, that is the combined effect of being in the ith row and the jth column. The α_i and β_j are referred to as the 'main effects' (for row and column factors, respectively) and the γ_{ij} as the non-additive interaction effects.

The interaction term represents the possibility that the whole is more than the sum of the parts, hence 'non-additive'. In the particular case of this experimental design, therefore, the parameters are μ, α_1, α_2, β_1, β_2, γ_{11}, γ_{12}, γ_{21}, and γ_{22}, that is nine parameters. However, apparently there are only four equations available: Equation (5.9) for η_{11}, η_{12}, η_{21}, and η_{22}. Therefore, additional constraints are necessary on the parameters. One set of constraints is simply to set some of the parameters of each type to 0:

$$\alpha_1 = 0$$

$$\beta_1 = 0$$

$$\gamma_{11} = 0$$

$$\gamma_{12} = 0$$

$$\gamma_{21} = 0$$

This now results in nine equations. In fact, this is exactly equivalent to the original formulation shown in Equation (5.8) with $\mu = \beta_{phys,0}$, $\alpha_2 = \beta_{phys,1}$, $\beta_2 = \beta_{phys,2}$, and $\gamma_{22} = \beta_{phys,3}$. The parameters set to 0 are said to be 'aliased' to 0. Hence, the results are always relative to these as 'base cases' – for example, the estimate for β_2 (or equivalently, $\beta_{phys,2}$) represents the effect of Gaze = InGroup relative to Gaze = OutGroup.

Here, we use another way to constrain the parameters, which is by setting the sum of each type to 0:

$$\alpha_1 + \alpha_2 = 0$$

$$\beta_1 + \beta_2 = 0$$

$$\gamma_{11} + \gamma_{12} = 0$$

$$\gamma_{21} + \gamma_{22} = 0$$

$$\gamma_{11} + \gamma_{21} = 0$$

$$\gamma_{12} + \gamma_{22} = 0$$

However, note that the last equation, for example, can be derived from the first three, since

$$\gamma_{11} + \gamma_{12} + \gamma_{21} + \gamma_{22} = 0$$

by summing the first two equations, but from the third equation, $\gamma_{21} = -\gamma_{11}$, and hence $\gamma_{12} + \gamma_{22} = 0$, which is the last equation. Hence, altogether there are five independent additional constraints, which then make up the nine required.

In general, suppose there are two factors, and the 'row' factor has p levels and the 'column' factor q levels. Then, the model is still Equation (5.9), but the constraints are

$$\sum_{i=1}^{p} \alpha_i = 0$$

$$\sum_{j=1}^{q} \beta_j = 0$$

$$\sum_{i=1}^{p} \gamma_{ij} = 0, j = 1,2, \ldots, q$$

$$\sum_{j=1}^{q} \gamma_{ij} = 0, i = 1,2, \ldots, p$$

In this set-up, there are $1 + p + q + pq$ parameters (μ, α_i, β_j, γ_{ij}). The number of equations is pq for the η_{ij}, 2 for the first two constraints, and $q + p - 1$ for the last two constraints (γ_{ij}) (as before one of the constraints on the γ_{ij} can be derived from the others). Hence, with these constraints, the number of parameters and the number of equations match.

Let's consider a bit more the meaning of this parameterization, which is usually called 'centred' or 'sum to zero constraints'. This is best understood if the link function were identity, in which case Equation (5.9) becomes

$$\mu_{ij} = \mu + \alpha_i + \beta_j + \gamma_{ij}$$
$$i = 1, \ldots, p; j = 1, \ldots q$$

where μ_{ij} represents the theoretical mean of the (i, j)th cell in the design. Taking the sum over j and dividing by q, we obtain

$$\bar{\mu}_{i.} = \mu + \alpha_i$$

so that $\alpha_i = \bar{\mu}_{i.} - \mu$, where $\bar{\mu}_{i.}$ is the mean of the ith row. Hence α_i can be interpreted as the specific effect of the ith level of the row factor, that is the difference between the mean of the ith level and the overall mean. Similarly $\beta_j = \bar{\mu}_{.j} - \mu$, representing the effect of the jth level of the column factor. Then the interaction term γ_{ij} represents the difference between the effect of being in the ith level of the row factor together with the jth level of the column factor and the effects of being in the ith level of the row factor plus the jth level of the column factor – that is, it is the effect of being in that cell over and above the separate additive effects of the factor levels. What this hopefully makes clear is that if there is no 'row effect' (i.e. the levels produce the same results in the response variable), then the α_i will be 0, and similarly for the 'column effect' and β_j. If there is no combined effect of the ith row level and the jth column effect over and above their additive effect, then the γ_{ij} will be 0.

Before moving on, it should be noted that the model expressed in Equation (5.9) can be represented in the same matrix notation as in Equation (5.4) – so that in the case of an identity link and normal likelihood function, the parameterization (Equation 5.9) is just a special case of the general linear model. In particular, let

$$\beta = \begin{bmatrix} \mu \\ \alpha_1 \\ \vdots \\ \alpha_p \\ \beta_1 \\ \vdots \\ \beta_q \\ \gamma_{11} \\ \vdots \\ \gamma_{pq} \end{bmatrix}$$

then the design matrix $\boldsymbol{X}$ consists of entries of 1 and 0 so that the multiplication of each row of $\boldsymbol{X}$ by the column vector $\boldsymbol{\beta}$ results in the required instance of Equation (5.9).

We show how our particular case with two binary factors Group and Gaze can be coded in Stan, as usual showing the differences from the previous program in red.

Box 5.4

```
data {
  int<lower=0> n;           //number in sample
  int<lower=0> k;           //number of parameters
  int g[n];                 //note - now treated as int
  int z[n];
  int<lower=0> yphys[n];              //num physical
  int<lower=0> yverb[n];              //num verbal
  vector<lower=0>[n] tphys;           //time physical
  vector<lower=0>[n] tverb;           //time verbal
}

transformed data {
  int r[n];
  int c[n];
```

```
  for(i in 1:n){
    r[i] = g[i] + 1; //Group condition 1 = OutGroup, 2 = InGroup
    c[i] = z[i] + 1; //gaze condition 1 = NoLookAt, 2 = LookAt
  }
}

parameters {
  real mu_phys;
  real alpha_physd;
  real beta_physd;
  real gamma_physd;
  real mu_verb;
  real alpha_verbd;
  real beta_verbd;
  real gamma_verbd;
  real<lower=0> phi1;
  real<lower=0> phi2;
}

transformed parameters {
  vector[2] alpha_phys;
  vector[2] beta_phys;
  matrix[2,2] gamma_phys;
  vector[2] alpha_verb;
  vector[2] beta_verb;
  matrix[2,2] gamma_verb;

  alpha_phys[1] = -alpha_physd;
  alpha_phys[2] = alpha_physd;

  beta_phys[1] = -beta_physd;
  beta_phys[2] = beta_physd;

  gamma_phys[2,2] = gamma_physd;
  gamma_phys[1,2] = -gamma_physd;
  gamma_phys[2,1] = -gamma_physd;
  gamma_phys[1,1] = gamma_physd;

  alpha_verb[1] = -alpha_verbd;
  alpha_verb[2] = alpha_verbd;
```

(Continued)

```
  beta_verb[1] = -beta_verbd;
  beta_verb[2] = beta_verbd;

  gamma_verb[2,2] = gamma_verbd;
  gamma_verb[1,2] = -gamma_verbd;
  gamma_verb[2,1] = -gamma_verbd;
  gamma_verb[1,1] = gamma_verbd;

}

model {
  real mu1;
  real mu2;
  real s = 10;

  mu_phys ~ normal(0,s);
  alpha_physd ~ normal(0,s);
  beta_physd ~ normal(0,s);
  gamma_physd ~ normal(0,s);

  mu_verb ~ normal(0,s);
  alpha_verbd ~ normal(0,s);
  beta_verbd ~ normal(0,s);
  gamma_verbd ~ normal(0,s);

  phi1 ~ gamma(2,0.1);
  phi2 ~ gamma(2,0.1);

  for(i in 1:n){
      mu1 = exp(mu_phys + alpha_phys[r[i]] + beta_phys[c[i]] +
          gamma_phys[r[i],c[i]]);
      mu2 = exp(mu_verb + alpha_verb[r[i]] + beta_verb[c[i]] +
          gamma_verb[r[i],c[i]]);
      yphys[i] ~ neg_binomial_2(mu1,phi1);
      yverb[i] ~ neg_binomial_2(mu2,phi2);
  }
}

generated quantities{
  int yphys_new[n];
  int yverb_new[n];
  vector[n] log_lik_phys;
  vector[n] log_lik_verb;
  real mu1;
  real mu2;
```

```
    for(i in 1:n){
      mu1 = exp(mu_phys + alpha_phys[r[i]] + beta_phys[c[i]] +
          gamma_phys[r[i],c[i]]);
      mu2 = exp(mu_verb + alpha_verb[r[i]] + beta_verb[c[i]] +
          gamma_verb[r[i],c[i]]);
      yphys_new[i] = neg_binomial_2_rng(mu1,phi1);
      yverb_new[i] = neg_binomial_2_rng(mu2,phi2);
      log_lik_phys[i] = neg_binomial_2_lpmf(yphys[i]|mu1,phi1);
      log_lik_verb[i] = neg_binomial_2_lpmf(yphys[i]|mu2,phi2);
    }
}
```

The 'transformed data' block allows us to create new variables from the original data. Recall that previously we had coded the factor levels as 0 and 1. In the current formulation, we need to code them as 1 and 2. This is because in the earlier formulation they were actual variables with possible values 0 or 1. Here, they are indices into the arrays for α, β, and γ. The transformed data block creates two new variables, 'r' and 'c', for rows and columns, respectively, to meet this requirement. This is because these parameters will be indices into vectors and matrices, which must be positive integers.

In this case since, for example, $\alpha_1 + \alpha_2 = 0$, we only need to declare α_2 as a parameter since $\alpha_1 = -\alpha_2$. For notational convenience, we call the parameter 'alpha_physd' (phys for the physical interventions) and the 'd' because we can think of this as α_2' ('d' for 'dash'). Similarly, we can do the same for the other parameters, treating the 'physical' and 'verbal' separately.

Then, in the 'transformed parameters' block, we define those parameters that we are really interested in, for example $\alpha_2 = \alpha_2'$ and $\alpha_1 = -\alpha_2'$, and similarly for each of the parameters.

In the 'model' block, we assign the *normal*(0,10) distribution as the prior for each of the originally defined parameters, and the 'transformed parameter' block automatically assigns corresponding distributions to the actual parameters of interest. Then, the models are defined as before. However, consider the meaning of 'alpha_phys[r[i]]'. The integer array value 'r[i]' is the Group level (1 for OutGroup or 2 for InGroup) for the *i*th participant. Hence, 'alpha_phys[r[i]]' corresponds (for the physical response) to α_1 or α_2 depending on whether the *i*th participant was in the OutGroup or InGroup level of the Group condition. Similarly, 'gamma_phys[r[i],c[i]]' corresponds to $\gamma_{r[i],c[i]}$, where 'r[i]' is the Group level (1 or 2) and 'c[i]' is the Gaze level (1 or 2) for the *i*th participant. Running through all the participants, the distributions for the parameters are updated as usual.

Table 5.10 shows the results for this version of the model. Recall that here the parameters have different meanings, and also the prior distributions will be different. In Table 5.9, each parameter represented the effect of the main or interaction effects of the Group = InGroup, Gaze = LookAt compared to a zero effect of the base cases Group = OutGroup and Gaze = NoLookAt. We discussed the meaning of the new parameters earlier. However, remember this is only a different parameterization – the underlying model is the same and overall the model will give the same results. For example, the predicted posterior values from this model will be the same as for the previous version (apart from small differences due to the simulations). To illustrate this, adding the parameter values $\mu + \alpha_2 + \beta_2 + \gamma_{22}$ from Table 5.10 results in 1.88 and adding the equivalent from Table 5.9, $\beta_{phys,0} + \beta_{phys,1} + \beta_{phys,2} + \beta_{phys,3}$, results in 1.87 (differences due to the different simulations and rounding error). Moreover, the qualitative conclusions from the two different parameterizations are the same.

Table 5.10 Summaries of posterior distributions of the parameters for the negative binomially distributed likelihood model for both numbers of physical and verbal interactions, using the alternative parameterization in Equation (5.9)

Parameter	Effect	*M*	*SD*	2.5%	97.5%	Prob > 0
numphysical						
μ		1.39	0.22	0.97	1.83	1.000
α_2	Group	0.54	0.22	0.09	0.97	0.989
β_2	Gaze	−0.14	0.22	−0.57	0.29	0.749
γ_{22}	Group × Gaze	0.09	0.22	−0.34	0.52	0.657
ϕ_{phys}		0.77	0.26	0.40	1.39	
numverbal						
μ		1.45	0.23	1.03	1.95	1.000
α_2	Group	0.35	0.24	−0.11	0.82	0.929
β_2	Gaze	−0.26	0.23	−0.72	0.20	0.122
γ_{22}	Group × Gaze	0.07	0.23	−0.38	0.52	0.627
ϕ_{verb}		0.70	0.23	0.36	1.23	

Note: Prob > 0 is the posterior probability that the parameter is > 0.

In most of what follows in this book, we will use the first parameterization rather than the second, only for pedagogical reasons; however, the central parameterization may sometimes be easier for interpretation. We have only shown this for binary factors (i.e. with two levels) in which case the representation in Stan is particularly easy. However, for the more general case, see Section 1.7 of the Stan User Guide https://mc-stan.org/docs/2_25/stan-users-guide/parameterizing-centered-vectors.html

Missing Data

We mentioned earlier that there were two sets of missing data. The two missing ones were one with Group = OutGroup and Gaze = LookAt, and the other was Group = OutGroup and Gaze = NoLookAt. This is evident from Table 5.5. In the models above, we took the simplest route to deal with these missing data – we just ignored the issue and deleted these individuals from the data set. Alternative methods involve some type of estimation of the missing values – for example, by setting each missing value to be the mean of other values under the same condition as the missing data, or basing the estimation on the values surrounding the missing value, and so on. A Bayesian approach is also based on estimation, but with a very simple idea: the missing data are unobservables, and can be treated as parameters, and dealt with almost identically to other parameters. Here, we take the opportunity to show how to include the timings in the analysis, specifically the amount of time participants engaged in a physical intervention: *timephysical*: t_{phys}. Second, we will show a simple example of how to estimate the two missing values on this variable.

Figure 5.10 shows the histograms of *timephysical* by the conditions. (We abbreviate this to *tphysical*.) What is clear is that this variable is highly unlikely to follow a normal distribution. As we noted in Chapter 2, time variables will often have an exponential or Gamma distribution. Here, we assume the more general Gamma form (Chapter 2). This has two parameters α and β. Hence, the likelihood is

$$t_{phys} \sim Gamma(\alpha, \beta) \tag{5.10}$$

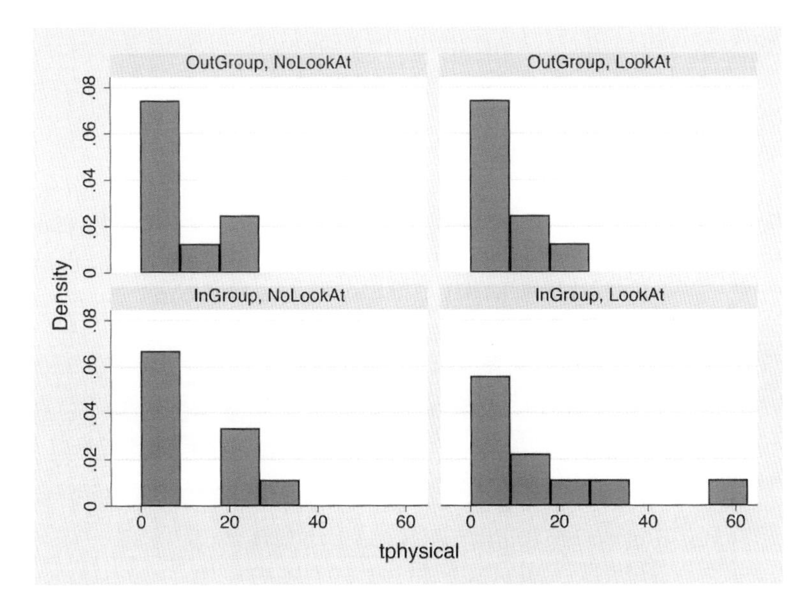

Figure 5.10 Histograms of timephysical by Group and Gaze

Moreover,

$$E(t_{phys} \mid \alpha, \beta) = \frac{\alpha}{\beta} \tag{5.11}$$

The linear predictor, for individual *i*, will be of the same form as Equation (5.8) with different link function and parameters:

$$\mu_i = \exp(\beta_{tphys,0} + \beta_{tphys,1} g_i + \beta_{tphys,2}\, z_i + \beta_{tphys,3}(g_i z_i))$$

The log link is used because it is required that the mean is positive. Hence, if we set

$$\alpha_i = \mu_i \beta$$

then

$$E(t_{phys,i}|\alpha_i, \beta) = \mu_i$$

as required.

Now to estimate the missing values on *tphysical*, we adopt the very simple assumption that the model remains unchanged for the missing data. Suppose that $t_{phys,m1}$ and $t_{phys,m2}$ are the two missing values. Then we assume that

$$t_{phys,m1} \sim Gamma(\alpha_{m1}, \beta)$$

$$t_{phys,m2} \sim Gamma(\alpha_{m2}, \beta)$$

where, for the first,

$$g_{m1} = 0, z_{m1} = 1 \text{ (OutGroup, LookAt)}$$

$$g_{m1} = 0, z_{m1} = 0 \text{ (OutGroup, NoLookAt)}$$

In the data block of the program, the variables are specified as before, except that now we include *tphys* for *timephysical*. We also include the number of missing values ('n_miss') and the factor values for those missing values ('gm', 'zm').

Box 5.5

```
data {
  int<lower=0> n;         //number in sample
  int<lower=0> k;         //number of parameters
  vector<lower=0,upper=1>[n] g;  //Group condition
  vector<lower=0,upper=1>[n] z;  //gaze condition
  int<lower=0> yphys[n];         //num physical
  vector<lower=0>[n] tphys;      //time physical
  int<lower=0> n_miss;           //number missing = 2
  vector<lower=0,upper=1>[n_miss] gm;  //Group condition missing
  vector<lower=0,upper=1>[n_miss] zm;  //gaze condition missing
}

parameters {
  vector[k] b_phys;
  real<lower=0> phi;
  vector[k] b_tp; //parameters for tphysical
  real<lower=0> beta;
  vector<lower=0>[n_miss] tp_miss; //missing values on physical responses
}
```

```
model {
  real mu;

  b_phys ~ normal(0,5);
  b_tp ~ normal(0,5);

  phi ~ gamma(2,0.1);
  beta ~ gamma(2,0.1);

  for(i in 1:n){
      //number physical
      mu = exp(b_phys[1] + b_phys[2]*g[i] + b_phys[3]*z[i] +
         b_phys[4]*g[i]*z[i]);
      yphys[i] ~ neg_binomial_2(mu,phi);

        //time physical
        mu = exp(b_tp[1] + b_tp[2]*g[i] + b_tp[3]*z[i] +
           b_tp[4]*g[i]*z[i]);
        tphys[i] ~ gamma(mu*beta,beta);
  }

  //deal with the missing values for tphys
  //assume the same model as for the observed values

  for(i in 1:n_miss){
     mu = exp(b_tp[1] + b_tp[2]*gm[i] + b_tp[3]*zm[i] +
        b_tp[4]*gm[i]*zm[i]);
     tp_miss[i] ~ gamma(mu*beta,beta);
  }
}
```

In the parameters block, we now add the parameters ('b_tp') corresponding to $\beta_{tphys,j}$ and the scale parameter for the Gamma distribution (beta) as in Equation (5.11). We also declare 'tp_miss' as the vector of (2) missing values.

In the model block, we add now the specifications for the tphys variable as in Equations (5.10) and (5.11), and we also use the same model for the specification of the missing values 'tp_miss'.

The results are shown in Table 5.11. The findings for the number of physical interventions are unchanged. The time of physical interventions does not add anything much to this, the probabilities of Group or the interaction term having an effect are moderate. However, there are estimates of the missing values, summarized in Figure 5.11. Although there are sharp modes for the posterior distributions, the variance is quite high in both cases.

Table 5.11 Summaries of posterior distributions of the parameters for the negative binomially distributed likelihood model for number and time of physical interactions

Parameter	Coefficient of	*M*	*SD*	2.5%	97.5%	Prob > 0
β_{phys_0}		1.07	0.47	0.22	2.06	0.994
β_{phys_1}	Group	0.91	0.62	–0.36	2.14	0.935
β_{phys_2}	Gaze	–0.44	0.66	–1.76	0.82	0.251
β_{phys_3}	Group × Gaze	0.34	0.88	–1.38	2.12	0.646
ϕ_{phys}		0.76	0.25	0.39	1.33	
β_{tp_0}		2.11	0.31	1.48	2.68	1.000
β_{tp_1}	Group	0.26	0.38	–0.47	1.02	0.768
β_{tp_2}	Gaze	–0.24	0.41	–1.08	0.54	0.281
β_{tp_3}	Group × Gaze	0.29	0.54	–0.74	1.35	0.709
β		0.09	0.02	0.06	0.14	
$t_{phys,m1}$		6.78	8.86	0.01	31.40	
$t_{phys,m2}$		8.73	10.86	0.01	38.35	

Note: Prob > 0 is the posterior probability that the parameter is > 0.

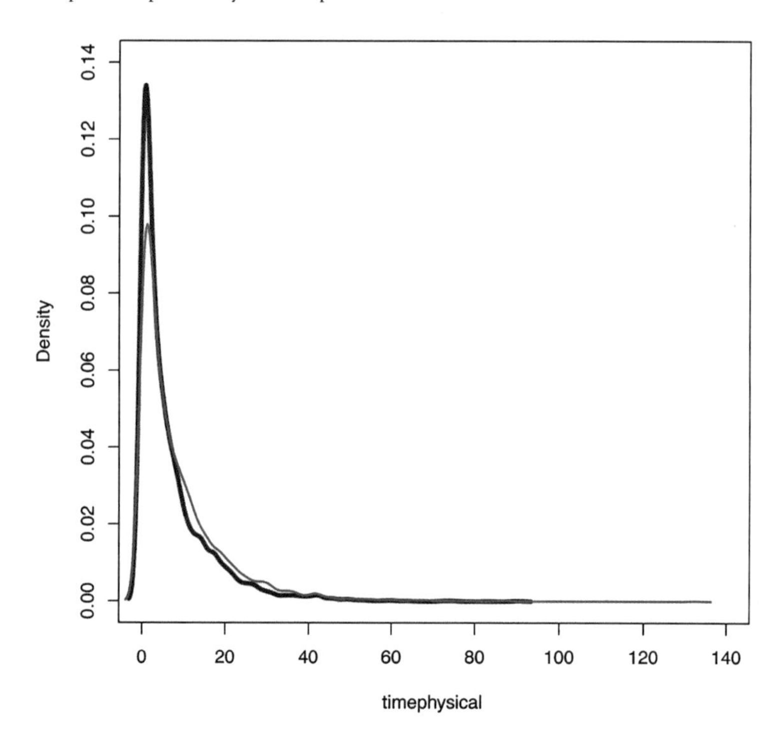

Figure 5.11 Posterior distributions for $t_{phys,\ m1}$ (black) and $t_{phys,\ m2}$ (blue)

Instead of using only the factor variables in the estimation of the missing data, we could have used other variables, such as various questionnaire scores (next chapter), in addition to these.

Summary

In this chapter, we have considered two fundamental general models – the general linear model, which restricts the mean of the normally distributed response variable to be equal to an expression that is linear in the parameters, and the generalized linear model, which applies to any distribution in the exponential family and which allows a non-identity link function between the mean and the linear model.

It is important to note that in the Bayesian method, we are far from restricted to the normal distribution, we do not have to follow canonical links, and we can apply all of these techniques to distributions that are not members of the exponential family. We have presented methods that connect well with classical statistics, but we are not bound by those models.

As an aside, we also briefly considered the issue of missing data. This does not introduce anything fundamentally new. The missing values are treated as parameters. Instead of giving the missing values the same model as the observed (i.e. relying on the same parameter distributions), we could adopt different strategies of making the distributions of the missing values dependent on different linear predictor combinations. For extensive discussions of missing values, see Little and Rubin (2019), and see the Stan user guide (https://mc-stan.org/docs/2_20/stan-users-guide/missing-data.html).

In the next chapter, we will consider a number of different models. The bystander experiment also included responses on questionnaires, and we will present methods for analysis of these.

Online Resources

The data, R, and Stan programs that go along with this chapter are available at

www.kaggle.com/melslater/slater-bayesian-statistics-5a
www.kaggle.com/melslater/slater-bayesian-statistics-5b

Six

QUESTIONNAIRES AND NON-QUANTITATIVE RESPONSES

Introduction

In the previous chapter we considered two general models, the general linear model and the generalized linear model. The general linear model is 'general' because it encompasses many different types of set-up as we saw. It is employed when there are situations where a response variable is considered as a function of multiple explanatory variables, or when there has been a specific experimental design that relates different levels of factors to variations in the response variable. The generalized linear model includes two enhancements: first, the response variable need not have a normal distribution but a distribution that is a member of the exponential family of distributions; and second, the relationship between the linear predictor and the mean of the response variable need not be one of identity. We considered examples concerned with reading age in relation to class size and family income, and another concerned with bystander behaviour in violent situations.

From the Bayesian perspective, what is important are the fundamental ideas. There is no requirement whatsoever to restrict these techniques to members of the exponential family, no necessity at all to use canonical link functions. In principle, any distribution appropriate to the situation at hand may be used and any link function. There is not even a requirement for the mean of the response variable to be linked to a function that is linear in the parameters. These requirements are based on classical statistical approaches, for mathematical convenience and computational tractability. In the Bayesian method, solutions to the integrals involved in finding the posterior probability distribution are carried out through simulation, so the issue of mathematical tractability is not pertinent. However, of course, some models may not have a solution (the simulations may not converge).

One thing common to all our examples so far is that the response variable has been numerical. It has been a theoretically continuous variable such as age at death in Chapter 4, or reading age in Chapter 5, or a discrete count variable such as the number of interventions by a bystander. The total time of interventions is also a continuous variable. In this chapter, we extend our models further by allowing for non-quantitative variables as responses. In particular, experimental studies or surveys of people typically include questionnaire responses. For example, in the bystander study, one question in the questionnaire that participants answered was 'After the argument started, I felt I should do something to stop it', where the possible responses were 1 = *Not at all* to 7 = *Very much so*. Participants were required to select a number from 1, 2, ..., 7 to indicate their level of agreement with this statement. This is an example of an ordinal scale – the responses are ordered. These responses are not numerical values. If they were numbers, we could say, for example, that a response of '6' is double that of '3', or that the difference between '7' and '5' has the same meaning as that between '3' and '1'. Instead, all we can say is that the responses are ordered, that '2' is greater than '1' but not that '2' is double '1', or that the distance between them is 1.

Nevertheless, many researchers do treat ordinal scale responses as if they were numbers and carry out operations such as finding their means and standard deviations, or even full analysis of variance with such responses treated as if they were actual numbers. It makes no sense to take the mean of people's responses on an ordinal questionnaire. Taking the mean or standard deviation assumes that the responses are additive – that is, two people with scores of '1' and

'3' together have the same mean response as someone who scored '2'. In this chapter, we will see how to analyse such response variables but without ever treating the values as quantitative.

Another type of response that is often included in questionnaires is categorical. This is where the response is just a category but with no implied order. For example, if a question asked, 'What is the country of your birth?', the answer would be one from a set, with no ordering implied.

The last type of response variable to be considered is a binary one. This is where there are only two possible values to a response variable that can always be coded as 0 or 1. For example, in the bystander study, did the participant move towards the protagonists or not?

In this chapter, we will consider each of these in turn, again first with some theory and then continue the bystander example from the previous chapter.

Ordered Regression Models

Concepts

Suppose we have responses to a question in a questionnaire which has k possible answers 1, 2, ..., k, where these are ordered responses. Let's denote the responses as observations on a random variable Y. As discussed above, we cannot treat these ordinal responses as if they were numbers. Instead, we suppose that there is an underlying continuous random variable Y^* where the observation $Y = j$ occurs when Y^* is in a specific interval. Let $c_1, c_2, \ldots, c_{k-1}$ be unknown parameters such that $c_1 < c_2 < \ldots < c_{k-1}$. Then this model supposes that

$$
\begin{array}{ll}
Y = 1 & \text{if } Y^* \leq c_1 \\
Y = 2 & \text{if } c_1 < Y^* \leq c_2 \\
Y = k - 1 & \text{if } c_{k-2} < Y^* \leq c_{k-1} \\
Y = k & \text{if } c_{k-1} < Y^*
\end{array}
$$

This can be written as

$$Y = j \text{ if } c_{j-1} < Y^* \leq c_j, \text{ where } c_0 = -\infty \text{ and } c_k = \infty \tag{6.1}$$

Of course, Y^* is unknown (it is a *latent variable* supposedly existing behind the scenes but not itself observable), and the c_j are unobservable parameters which can be given prior distributions, and information about their values can be obtained from their posterior distributions. The idea is that whenever the 'true' unobservable Y^* crosses a threshold, then the *observed* ordinal response Y changes as a result. The c_j are sometimes referred to as the 'cut points'.

The logit function, as we have seen before, is defined as

$$\text{logit}(p) = \log\left(\frac{p}{1-p}\right) \tag{6.2}$$

where p is the probability of an event. Then logit(p) is the log odds of the event happening compared to it not happening.

If we write $\log\left(\frac{p}{1-p}\right) = u$ and solve for p, then

$$p = \frac{1}{1+e^{-u}} \equiv \text{logit}^{-1}(u) \tag{6.3}$$

As we saw in Chapter 2 (Equation 2.12), this is the cumulative distribution function on $-\infty < u < \infty$ of the logistic distribution (mean 0): $\text{logit}^{-1}(-\infty) = 0$ and $\text{logit}^{-1}(\infty) = 1$.

We consider the situation where there are n independent random variables which are observations on an ordinal variable Y, that is Y_i, $i = 1, 2,\ldots, n$, where each observation is an ordinal score 1, 2, ..., k. Suppose there are p independent or explanatory variables $x_1, x_2,\ldots, x_p$. As in the previous chapter, the linear predictor is

$$\eta_i = \sum_{s=1}^{p} \beta_s x_{si},\ i = 1, 2,\ldots, n$$

The proportional odds logistic model states that

$$\begin{aligned} &\text{logit}(P(Y_i \leq j)) = \alpha_j - \eta_i \\ &j = 1, 2,\ldots, k-1 \end{aligned} \tag{6.4}$$

Note that this model says that the log of the odds of a score at most j against a score of greater than j is a function of an intercept term α_j and the linear predictor. It is only the intercept terms that change for different values of j. Hence, for different values of j, the model implies a set of parallel lines (or hyperplanes) with different intercepts. This is an assumption of the model that needs to be checked. The intercept terms α_j are the values of the log odds when all the independent variables are 0, or all the $\beta_s = 0$ (i.e. where the independent variables do not influence the response).

Using Equation (6.3), we can write from Equation (6.4),

$$P(Y_i \leq j) = \frac{1}{1+e^{-(\alpha_j - \eta_i)}} \tag{6.5}$$

Note that this implies

$$P(Y_i = j) = \frac{1}{1+e^{-(\alpha_j - \eta_i)}} - \frac{1}{1+e^{-(\alpha_{j-1} - \eta_i)}}$$

Now consider the implications for the latent variable Y^*. From Equation (6.1),

$$P(Y_i \leq j) = P(Y^* \leq c_j)$$

Suppose Y_i^* has a logistic distribution with mean μ_i. Then,

$$P\left(Y_i^* \le c_j\right) = \frac{1}{1+e^{-\left(c_j-\mu_i\right)}} \quad (6.6)$$

Comparing Equations (6.5) and (6.6) if we put

$$\alpha_j = c_j$$

and

$$\mu_i = \eta_i \quad (6.7)$$

then we have the following explanation of the model for ordered logistic analysis. Each independent Y_i^* has a logistic distribution. The random variables denoting the observed ordinal responses Y_i follow the conditions given in Equation (6.1): whenever Y_i^* crosses a threshold c_j, the observed Y_i changes. The relationship with the independent variables is given by Equation (6.4), equivalently Equation (6.5) bearing in mind Equation (6.7). The c_j and β_s are the unobservable parameters. Hence, they will be given prior distributions, and posterior distributions can be found conditional on the observed data.

The Bystander Example Continued

Here we continue the bystander data example from Chapter 5. After participants had experienced the virtual reality scenario, they were given a questionnaire to answer. Amongst the questions were those shown in Table 6.1.

Table 6.1 Questions answered by participants in the bystander study

Question	Variable name	Short name
After the argument started, the aggressor was aware of me looking at him.	AggressorAware	aa
After the argument started, the victim looked at me wanting help.	VictimLooked	vl
After the argument started I looked around for help.	HelpMe	hm
After the argument started I was sometimes concerned for my own safety.	OwnSafety	os

Note: Each question was scored on a 1 to 7 scale, where 1 indicates strong disagreement and 7 indicates strong agreement with the corresponding statement.

The experimenters wanted to know how the responses to the questionnaire varied with the conditions of the experiment. Recall that there were two factors: Group (OutGroup = 0, InGroup = 1)

and Gaze (NoLookAt = 0, LookAt = 1), and the question is whether the questionnaire responses varied with these. In particular, here we consider the response HelpMe (hm).

The box plots for HelpMe are shown in Figure 6.1. It seems that InGroup may generally have higher scores than OutGroup, and possibly LookAt higher than NoLookAt. To examine this using a Bayesian analysis, we use Stan which has a built-in function for the ordered logistic model.

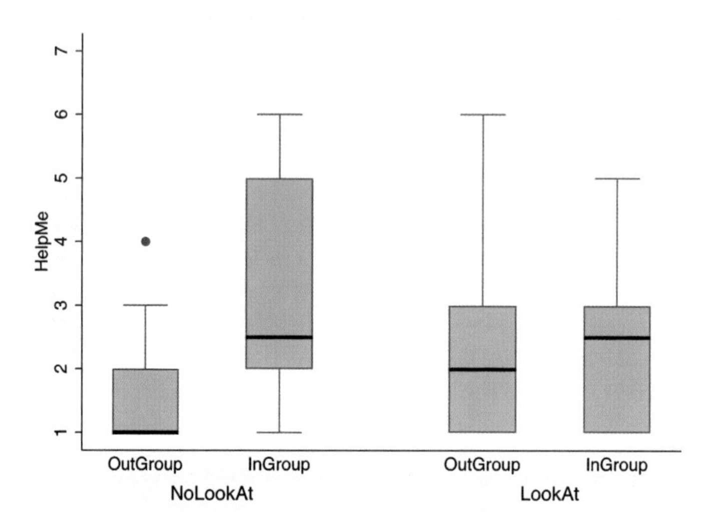

Figure 6.1 Box plots for HelpMe. The thick black horizontal lines are the medians and the boxes are the interquartile ranges

Box 6.1

```
data {
    int<lower=0> n;         //number in sample
    int<lower=0> p;         //number of parameters
    vector<lower=0,upper=1>[n] g;  //Group condition
    vector<lower=0,upper=1>[n] z;  //gaze condition
    int<lower=0> yphys[n];         //num physical
    int<lower=0> yverb[n];         //num verbal
    vector<lower=0>[n] tphys;      //time physical
    vector<lower=0>[n] tverb;      //time verbal
    int<lower=2> k;                //maximum ordinal response
    int<lower=1,upper=k> aa[n];       //aggressoraware
    int<lower=1,upper=k> vl[n];       //victimlooked
    int<lower=1,upper=k> hm[n];       //helpme
    int<lower=1,upper=k> os[n];       //ownsafety
}
```

```
parameters {
    vector[p] b_phys;
    vector[p] b_verb;
    real<lower=0> phi1;
    real<lower=0> phi2;
    ordered[k-1] c;
    vector[p-1] b_hm;
}

model {
  real mu1;
  real mu2;

  for(i in 1:p) {
      b_phys[i] ~ normal(0,10);
      b_verb[i] ~ normal(0,10);
  }
  //note that the above loop can be replaced by
  //b_phys ~ normal(0,10);
  //b_verb ~ normal(0,10);
  //and the same for the other parameters below
  phi1 ~ gamma(2,0.1);
  phi2 ~ gamma(2,0.1);

  for(i in 1:p-1){
      b_hm[i] ~ normal(0,10);
  }
  for(i in 1:k-1){
    c[i] ~ normal(0,10);
  }

  for(i in 1:n){
        mu1 = exp(b_phys[1] + b_phys[2]*g[i] + b_phys[3]*z[i] +
                   b_phys[4]*g[i]*z[i]);
        mu2 = exp(b_verb[1] + b_verb[2]*g[i] + b_verb[3]*z[i] +
                   b_verb[4]*g[i]*z[i]);
        yphys[i] ~ neg_binomial_2(mu1,phi1);
        yverb[i] ~ neg_binomial_2(mu2,phi2);
        hm[i] ~ ordered_logistic(b_hm[1]*g[i] + b_hm[2]*z[i] +
                                       b_hm[3]*g[i]*z[i],c);
  }
}
```

In the program above, we highlight the changes compared to the previous chapter. We include both the number of physical responses of the participant during the scenario, the same as in

Chapter 5, and the new response variable which is the questionnaire score HelpMe. This is so that the reader can see how multiple response variables can be included in the same overall model. Note the following:

The data block: we use now p for the number of parameters in the model (previously it was k). k is now the number of ordinal responses possible (k = 7). We define the new questionnaire variables.

The parameters block: The critical new entry is for c. This is declared with an *ordered* type. As we know the cut points c must be in increasing order. The 'ordered' declaration ensures this. Any prior distributions we declare over c will respect the ordering.

The model bock: The coefficients 'b_hm' are the parameter values for the linear predictor for the response variable 'hm'. We give both these and the cut points c the normal distribution with mean 0 and standard deviation 10. This means that their prior 95% credible intervals are in the range −20 to 20. We declare the response variable 'hm' to have the ordered_logistic distribution. This has the linear predictor and c as its parameters.

Table 6.2 Summaries of posterior distributions of the parameters for the ordered logistic model

Parameter	Coefficient of	Mean	*SD*	2.5%	97.5%	Prob > 0
yphys						
β_{phys_0}		1.08	0.47	0.21	2.05	0.993
β_{phys_1}	Group	0.90	0.63	−0.32	2.15	0.928
β_{phys_2}	Gaze	−0.46	0.70	−1.81	0.93	0.244
β_{phys_3}	Group × Gaze	0.35	0.92	−1.45	2.15	0.654
hm						
β_{hm_1}	Group	1.90	0.93	0.15	3.80	0.986
β_{hm_2}	Gaze	1.33	0.94	−0.42	3.24	0.925
β_{hm_3}	Group × Gaze	−2.09	1.24	−4.60	0.29	0.046
cut points						
c_1		0.59	0.72	−0.71	2.05	
c_2		1.48	0.74	0.13	3.00	
c_3		2.45	0.78	1.00	4.04	
c_4		3.12	0.83	1.57	4.85	
c_5		4.62	1.04	2.74	6.79	
c_6		12.14	5.06	5.26	24.39	

Note: Prob > 0 is the posterior probability that the parameter is > 0.

The results are shown in Table 6.2. For the number of physical responses the results are the same as in Chapter 5, as expected. Recall that the slightly different values are because Stan finds the posterior distributions through simulation. For the HelpMe variable, it seems that Group (InGroup) is associated with greater values (probability 0.986). Similarly with Gaze (probability = 0.925). However, the interaction effect, when there is InGroup and LookAt is associated with a decrease in score (probability = 1 – 0.046 = 0.954) – that is, less than the sum of the InGroup and LookAt effects.

We can easily see the proportional odds model at work. If we consider Equation (6.4), it is easier to understand when there is a single independent variable x. Then the equation would be (recalling that $\alpha_j = c_j$)

$$\text{logit}(P(Y_i \leq j)) = c_j - \beta x_i$$

Hence, the lines for different levels of the response j differ only by the intercept, they are all parallel. Note also that the interpretation of β becomes clear. Higher values of β mean lower log odds – that is, in favour of $P(Y_i > j)$. So higher values of β are associated with higher questionnaire score values.

From Equation (6.4), it should be clear that the difference between two logit values for two different values of an independent variable should be the same independently of j. We can easily check this in the Stan program (see Online Resources in the section with comment '#checking the proportional odds assumption'). We can find the posterior distributions of all the $\text{logit}(P(Y_i \leq j))$ and then find the means of these distributions under the constraints, for example that Group = 1 and Group = 0. Whatever the value of j, the difference between these is always –0.85. Similarly, we can find the values for Gaze = 1 and Gaze = 0, and the difference between the means of the posterior distributions of the logits is always –0.225 for all j. No matter which differences we choose to test, we will always find the same result. The 'parallel lines' or 'proportional odds' assumption is built into the method.

Categorical Logistic Regression

Concepts

Here we consider the situation of a response variable that is categorical rather than ordinal. In other words, the responses are one of a set of categories, with no particular ordering relationship amongst them. For example, the Borgonian Armamentarium United football club might be interested in fans evaluating which type of shirt the club players should wear and assessing whether there are differences in choices between different genders or age groups. Here the 'type of shirt' would be the response variable, and gender and age would be the independent variables. There is no ordering amongst the possible values of the response variable, it is just a set of categories.

A categorical variable Y has a fixed set of r responses denoted as 1, 2, ..., r (the labelling is arbitrary). One of these is chosen as the 'root' – usually the rth one (and since the labelling is arbitrary, this could be any one of the actual categories). Then the model is

$$\log\left(\frac{P(Y_i = j)}{P(Y_i = r)}\right) = \eta_{ij} = \beta_{j0} + \beta_{j1}x_{1i} + \beta_{j2}x_{2i} + \ldots + \beta_{jk}x_{ki}$$

where

$$i = 1, 2, \ldots, n \tag{6.8}$$

and

$$j = 1, 2, \ldots, r$$

This expresses the log odds of category j against category r as the linear predictor. Note that, for each category, there is a different set of parameters β. When $j = r$, the left-hand side is 0, and therefore

$$\eta_{ir} = 0$$

meaning that all the corresponding $\beta_r = 0$.

Another way to write Equation (6.8) is

$$\frac{P(Y_i = j)}{P(Y_i = r)} = e^{\eta_{ij}}$$

or

$$P(Y_i = j) = P(Y_i = r)e^{\eta ij}$$

Since $\sum_{s=1}^{r} P(Y_i = s) = 1$

$$P(Y_i = r)\sum_{s=1}^{r} e^{\eta_{is}} = 1$$

so that

$$P(Y_i = r) = \frac{1}{1 + \sum_{s=1}^{r-1} e^{\eta_{is}}}$$

leading to

$$P(Y_i = j) = \frac{e^{\eta_{ij}}}{1 + \sum_{s=1}^{r-1} e^{\eta_{is}}} \tag{6.9}$$

If z is a vector of length k, then the function 'softmax' is defined as

$$softmax(z) = \frac{e^z}{\sum_{j=1}^{k} e^{z_j}}$$

where e^z is a vector with ith element e^{z_i}. Hence, the vector of all probabilities (Equation 6.9) $P(Y_i = j)$, $j = 1, 2, \ldots, r$, can be written as

$$softmax(\eta_i) \tag{6.10}$$

where

$$\eta_i = \begin{pmatrix} \eta_{i1} \\ \eta_{i2} \\ \vdots \\ \eta_{ir} \end{pmatrix}$$

Recall that $e^{\eta_{ir}} = 1$ for all i.

In Chapter 5, we showed how to write the model using matrix form – see Equation (5.4). In order to understand the formulation needed in Stan, we also make use of that here. Equation (6.8) can be rewritten as

$$\begin{pmatrix} 1 & x_{11} & x_{21} \cdots & x_{k1} \\ 1 & x_{12} & x_{22} \cdots & x_{k2} \\ 1 & x_{13} & x_{23} \cdots & x_{k3} \\ \vdots & \vdots & \vdots \quad \vdots & \vdots \\ 1 & x_{1n} & x_{2n} \cdots & x_{kn} \end{pmatrix} \begin{pmatrix} \beta_{10} & \beta_{20} & \cdots & \beta_{r0} \\ \beta_{11} & \beta_{21} & \cdots & \beta_{r1} \\ \beta_{12} & \beta_{22} & \cdots & \beta_{r2} \\ \vdots & \vdots & \vdots & \vdots \\ \beta_{1k} & \beta_{2k} & \cdots & \beta_{rk} \end{pmatrix} \tag{6.11}$$

The (i, j)th entry in the product matrix is the right-hand side expression of Equation (6.8) – that is, η_{ij}. Or the ith row of the product matrix is

$$\eta_i' = (\eta_{i1}, \eta_{i2}, \ldots, \eta_{ir})$$

that is, the transpose of η_i in Equation (6.10).

A categorical response variable (Y) that has r possible outcomes has probabilities

$$P(Y = j) = p_j$$

$$j = 1, 2, \ldots, r$$

where $\sum_{j=1}^{r} p_j = 1$.

This is the categorical distribution. For the categorical logistic model, we simply have

$$p = softmax(\eta) \tag{6.12}$$

where p is the vector of probabilities and

$$\eta = \begin{pmatrix} \eta_1 \\ \eta_2 \\ \vdots \\ \eta_r \end{pmatrix} \text{ (we have omitted the } i \text{ for convenience)}$$

Note that as in the case of the ordered logistic model, Equation (6.12) ensures that all $p_j \geq 0$.

Categorical Logistic Regression Example

In the bystander experiment, the types of intervention that participants made were recorded and categorized as (1) speaking, (2) moving towards or reaching out to the protagonists, (3) looking around as if for help, or (4) no intervention. For each participant, we find the category that occurred the most often. We treat this response as a categorical variable, itype (intervention type), with $r = 4$ possible outcomes. This is obviously a categorical variable – there is no intrinsic ordering for these responses. The purpose of the analysis is to examine, as before, whether the independent factors, Group and Gaze, may influence this response. The Stan program is shown below.

Box 6.2

```
data {
    int<lower=0> n;         //number in sample
    int<lower=0> p;         //number of parameters
    vector<lower=0,upper=1>[n] g;  //Group condition
    vector<lower=0,upper=1>[n] z;  //gaze condition
    int r; //number of categories in itype (r=4)
    int<lower=1,upper=r> itype[n];
}

transformed data{
    vector[p] zeros = rep_vector(0,p);
    matrix[n,p] x;

    //create the X matrix
    x[1:n,1] = rep_vector(1,n);
    x[1:n,2] = g;
    x[1:n,3] = z;
    x[1:n,4] = g .* z;
}

parameters {
    matrix[p,r-1] b_it_raw;
}

transformed parameters{
    matrix[p,r] b_it;
    matrix[n,r] x_beta;

    b_it = append_col(b_it_raw,zeros);
```

```
    x_beta = x * b_it;
}

model {

  to_vector(b_it_raw) ~ normal(0, 10); //converts the matrix to a
                                        //column vector

  for(i in 1:n){
    itype[i] ~ categorical(softmax(to_vector(x_beta[i])));
    //to_vector converts row to column vector
  }
}

generated quantities{
  int itype_new[n];
  vector[n] log_lik_itype;
  vector[r] prob[n];

  for(i in 1:n){
      itype_new[i] = categorical_rng(softmax(to_vector(x_beta[i])));
      //to_vector converts row to column vector
      log_lik_itype[i] =
            categorical_lpmf(itype[i]|softmax(to_vector(x_beta[i])));
      prob[i] = softmax(to_vector(x_beta[i]));
  }
}
```

The 'data' block includes the new variable itype. Here $r = 4$ is the number of categories, so itype is bound between 1 and 4.

The 'transformed data' block as we saw in the previous chapter is where transformations of the data can be carried out, to produce new variables that are then available in the remainder of the program. In this case, for reasons that will be clear later, we construct the X matrix corresponding to the independent variables, as in Equation (6.11). This is done on a column-by-column basis. The new matrix is 'x', and 'x[1:n,j]' refers to the jth column of 'x'. So the first column is set to all 1, 'rep_vector(1,n)' produces a vector of length n where each entry is 1. Vectors in Stan are by default column vectors. The second column 'x[1:n,2]' is set to 'g' (Group), the third to 'z' (Gaze), and the fourth column to the interaction term, the element-wise product of 'g' and 'z'.

Now we want to make the matrix of βs from Equation (6.11), but recalling that by construction of the model $\beta_{rs} = 0$, $s = 1, \ldots, p$. This is done in two parts. In this 'parameters' block, we define 'b_it_raw' to be a matrix consisting of the first $r - 1$ columns of the matrix of βs. The 'transformed parameters' block is the parameter equivalent of 'transformed data' – it is where new parameters may be defined based on the parameters defined in the 'parameters' block. Here we make a new matrix of βs consisting of the first $r - 1$ columns of 'b_it_raw', and the final column of zeros. We defined that final column of zeros in 'transformed data', named zeros, and it is defined with 'rep_vector(0,p)' (a vector of length p with all 0s). Then 'append_col' appends the vector of zeros as the last column of 'b_it'. Then we construct the product matrix (Equation 6.11) as 'x_beta'.

Now consider the 'model' block. The 'to_vector' function transforms the matrix into a vector, and every element has a *normal*(0,10) distribution as prior. This is just a more convenient (and efficient) way instead of writing a 'for' loop that runs through each element of 'b_it_raw' to give it a prior distribution. Then itype has the categorical distribution, and the parameter is exactly as we defined above in Equations (6.11) and (6.12). Since 'x_beta' is a matrix, 'x_beta[i]' is the ith row. Here the 'to_vector' transposes it to be a column matrix as required by the function. Although for pedagogical reasons we have written it in this form, in fact a simpler and more efficient way would have been to write

```
itype[i] ~ categorical_logit(to_vector(x_beta[i]));
```

which means exactly the same.

The 'generated quantities' block computes new randomly generated values of predicted itype and also the quantity needed for the 'loo' analysis. It also includes the actual probabilities corresponding to Equation (6.12).

Table 6.3 shows the frequencies of category responses. 'Spoke' was the category with the most frequent response. Figure 6.2 shows the frequencies of occurrences of the categories by the Group and Gaze factor levels. For the Spoke category, the greatest frequency was for InGroup compared to OutGroup and LookAt is lower than NoLookAt in both Group conditions. Moreover, the InGroup frequency is the greatest of all frequencies in all categories. For the Moved category, we can see that the frequency is 0 for the (OutGroup, LookAt) condition. For the Looked category, by far the highest frequency is for the (Outgroup, LookAt) condition. For the No Interaction category, there appears to be no difference between OutGroup and InGroup, but LookAt has the highest frequency. To unravel all this, we can consider the results of the Bayesian model.

Table 6.3 Frequency table for itype

Category	Spoke	Moved	Looked	No Intervention
Frequency	16	8	6	8

Table 6.4 shows the summaries of the posterior distributions of the βs. Remember that these are relative to the No Interaction condition, which was set to zero. For the Spoke category, there is a probability of $1 - 0.129 = 0.871$ that Gaze (InGroup) is associated with a lower response. For the Moved condition, we see from Figure 6.2 that (OutGroup, LookAt) is 0, but (InGroup, LookAt) is non-zero. This is reflected in the negative coefficient for the main effect of Gaze and the positive coefficient for the interaction.

For the Looked category, we see the negative effect of the interaction, reflecting the 0 frequency for (InGroup, LookAt).

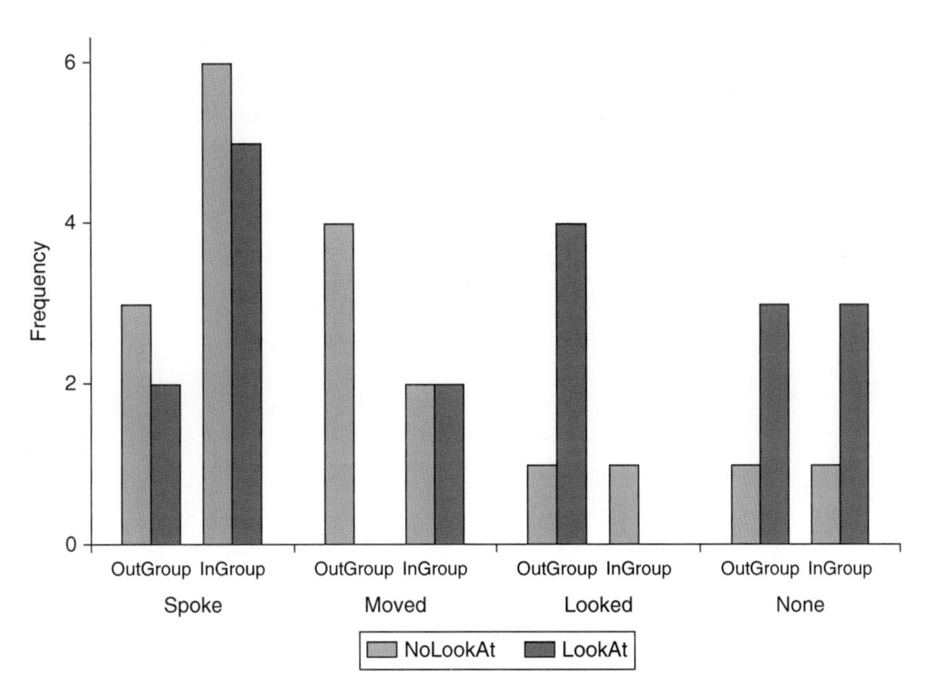

Figure 6.2 Bar chart showing the frequencies of the itype response by Group and Gaze

Table 6.4 Summaries of posterior distributions of the parameters for the categorical logistic model

Parameter	Coefficient of	Mean	*SD*	2.5%	97.5%	Prob > 0
Spoke						
β_{10}		1.28	1.27	−1.04	4.10	0.852
β_{11}	Group	0.92	1.76	−2.46	4.47	0.703
β_{12}	Gaze	−1.72	1.59	−5.05	1.18	0.129
β_{13}	Group × Gaze	0.11	2.14	−4.21	4.27	0.538
Moved						
β_{20}		1.58	1.22	−0.52	4.30	0.922
β_{21}	Group	−0.59	1.80	−4.19	2.88	0.352
β_{22}	Gaze	−7.68	4.06	−17.12	−1.69	0.002
β_{23}	Group × Gaze	6.18	4.28	−0.79	15.78	0.958
Looked						
β_{30}		−0.04	1.54	−3.09	3.05	0.497
β_{31}	Group	−0.17	2.24	−4.75	4.25	0.465
β_{32}	Gaze	0.37	1.73	−3.13	3.82	0.582
β_{33}	Group × Gaze	−8.96	6.02	−23.08	0.29	0.032

Note: Prob > 0 is the posterior probability that the parameter is > 0. The categories are Spoke, Moved, Looked, and No Intervention. The No Intervention category has all parameters aliased to 0.

We can perhaps gain a better understanding from the predicted posterior probability distributions ('probs' in the 'generated quantities' block in the program). Let the probabilities (Equation 6.12) be p_{Spoke}, p_{Moved}, p_{Looked}, and p_{None}. Figure 6.3 shows the density estimates for the predicted posterior distributions. Note that the means of these distributions are as follows:

$E(p_{Spoke}) = 0.421$

$E(p_{Moved}) = 0.21$

$E(p_{Looked}) = 0.158$

$E(p_{None}) = 0.211$

and these sum to 1 as required.

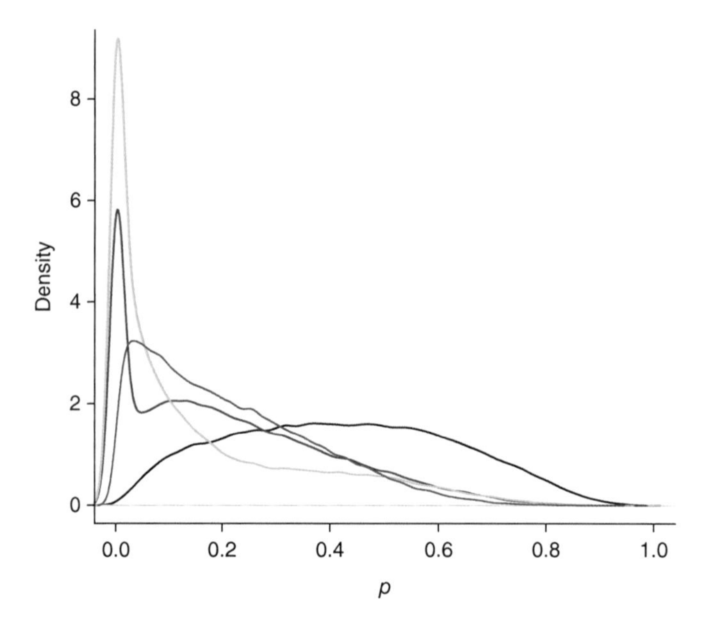

Figure 6.3 Posterior predicted distributions for the probabilities p_{Spoke} (black), p_{Moved} (red), p_{Looked} (green), and p_{None} (blue)

It is clear that the greatest probability is that a response will be in the Spoke category. For example, from these distributions we can find that

$P(p_{Spoke} > 0.5) = 0.369$

$P(p_{Moved} > 0.5) = 0.087$

$P(p_{Looked} > 0.5) = 0.082$

$P(p_{None} > 0.5) = 0.054$

There is more than four times probability that $p_{Spoke} > 0.5$ than each one of the others. Spoke being the most likely outcome corresponds with the frequencies in Table 6.3.

Next we can look at some of the posterior predicted distributions corresponding to the different levels of the factors Group and Gaze. For example, from Figure 6.2, it seems that (InGroup, NoLookAt) has the highest frequency. How does this turn out when we examine some of the posterior distributions?

Figure 6.4 shows that the distribution for Spoke in the condition Group = InGroup and Gaze = NoLookAt has the highest range of probabilities, reflecting what can be seen in Figure 6.2. For example, we can find that

$$P(p_{Spoke} > 0.7|\text{InGroup, NoLookAt}) = 0.269$$

$$P(p_{Spoke} > 0.7|\text{InGroup, LookAt}) = 0.107$$

$$P(p_{Spoke} > 0.7|\text{OutGroup, NoLookAt}) = 0.042$$

$$P(p_{Spoke} > 0.7|\text{OutGroup, NoLookAt}) = 0.010$$

The results therefore suggest that the most likely response is Spoke and also that this is most likely to occur in the Group = InGroup and Gaze = NoLookAt conditions. However, is it Group that is more important or Gaze?

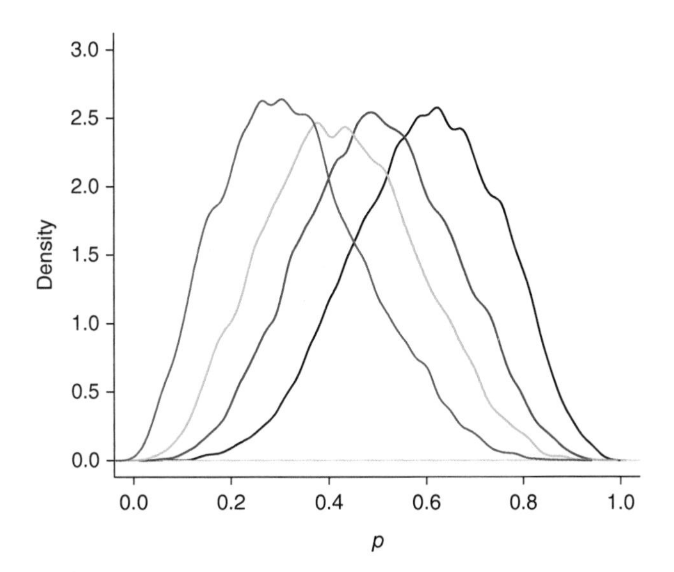

Figure 6.4 Posterior predicted distributions for the probability distributions: $f(p_{Spoke}|\text{InGroup, NoLookAt})$ (black), $f(p_{Spoke}|\text{InGroup, LookAt})$ (red), $f(p_{Spoke}|\text{OutGroup, NoLookAt})$ (green), $f(p_{Spoke}|\text{OutGroup, NoLookAt})$ and (blue)

Figure 6.5 shows the posterior predicted distributions comparing the two Group conditions and the two Gaze conditions. Figure 6.5A shows a wide separation between the Group conditions with the InGroup having greater probability to respond with Spoke than the OutGroup. However, there is little difference between the two Gaze conditions. Overall, the evidence suggests that the Spoke response was most probable response, and especially for those in the InGroup condition. There is some evidence, but not so strong, that the (InGroup, NoLookAt) condition was associated with the greatest Spoke response.

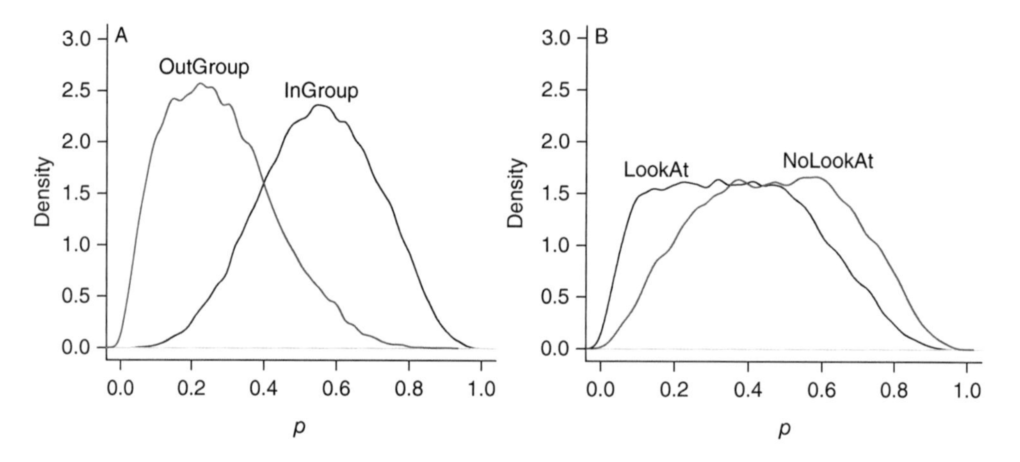

Figure 6.5 Posterior predicted distributions for the probability distributions: (A) $f(p_{Spoke}|\text{InGroup})$ (black) and $f(p_{Spoke}|\text{OutGroup})$ (red) and (B) $f(p_{Spoke}|\text{LookAt})$ (black) and $f(p_{Spoke}|\text{NoLookAt})$ (red)

Criticizing the Model

Having constructed this model, we should criticize it. How well does the posterior predicted distribution conform with the observed values? Recall that for each individual *i*, we obtain a probability distribution for the predicted itype, that in the program ('generated quantities') we call 'itype_new'. For example, from the original data, we know that individual 1 did not intervene, and therefore the score for itype is 4. Figure 6.6 shows the distribution for this individual, which has mode of 3. However, the probability that itype = 3 for individual 1 is reasonably high as 0.322. It is important not to conclude that the modal prediction of 3 is quite close to the actual value of 4. Remember, for categorical responses '3' and '4' are not numbers. 'Looked' is no more close to 'None' than any of the other categories.

We can compute the posterior predicted probability for each individual's actual itype score, that is 'p[i] = Prob(itype_new[i] = itype[i])' for each *i*.

Figure 6.7 shows the result of this computation. From Figure 6.7A, it seems that the predictions are not too good, they appear to be better for Spoke than the others. What we would like to see for a good model is a horizontal band with probability at least 0.5 across all the categories. Figure 6.7B shows the histogram of all of these probabilities. What we would like to see is the probabilities bunched towards higher levels (at least 0.5). However, we do not see these desired results. So overall, we would conclude that the model requires revision – the logit assumption may not be appropriate.

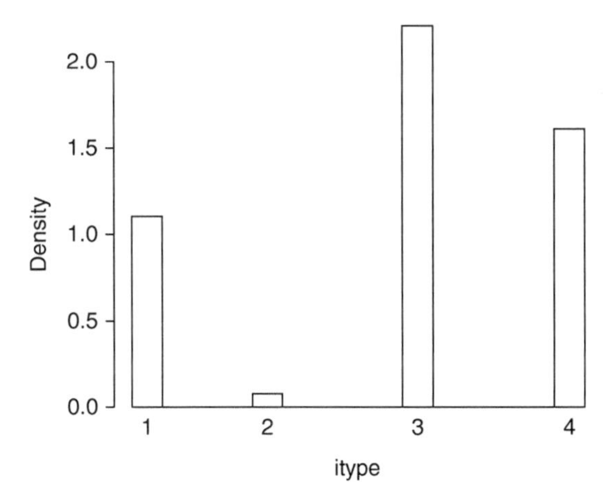

Figure 6.6 Predicted posterior distribution for itype for individual 1

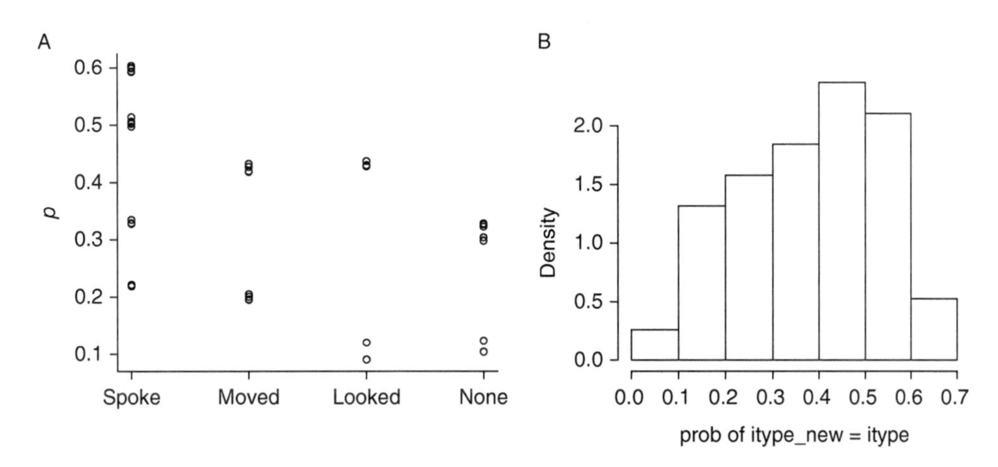

Figure 6.7 Predicted posterior probabilities that the predicted type ('itype_new') is the same as the observed type (itype): (A) the probabilities by category and (B) the histogram of probabilities

The 'leave-one-out' ('loo') analysis is also not good, and 32/38 Pareto *k* values give 'good' results, 3 are 'ok', but 3 are 'bad'. The elpd_loo = –56.6 ± 5.8 (*SE*), and the effective number of parameters p_loo = 14.7, which is OK given that the true number of parameters is 12.

The 'bad' Pareto *k* values mean that some observations are particularly influential and, possibly, biasing the results. This is like when in regression all the observations fall approximately on a straight line, but some of the observations are far away at one end of the line, thus pushing the whole regression line up or down away from the bulk of the observations. In order to see the impact of the 'bad' observations, we can rerun the model with these eliminated and again look at the 'loo' results. This highlighted one further 'bad' observation. Therefore, four observations were eliminated, and the model run again, and this time all the Pareto k values were 'good'. In the end, this makes hardly any difference to the graphs above, but it does make a difference to the results in Table 6.4.

Table 6.5 shows the new results, having eliminated the four problematic observations. For these reduced data, we can be more confident in the results. However, now we have lost some of the observations, so it is at the end of the day not clear whether these new results are reliable, since the 'problematic' observations may have indicated important departures from the model, but which can now not be captured in the reduced data model. So it is a matter of going back to the original data, checking everything, examining whether the four problematic data points reflect individuals who are somehow different from the others, perhaps indicating that other explanatory variables need to be included. For example, age of the individual might play an important role, but we have not tested this here.

Table 6.5 Summaries of posterior distributions of the parameters for the categorical logistic model after removing four observations

Parameter	Coefficient of	Mean	SD	2.5%	97.5%	Prob > 0
Spoke						
β_{10}		5.28	3.04	0.55	12.27	0.904
β_{11}	Group	4.12	4.27	−3.59	13.07	0.967
β_{12}	Gaze	−5.74	3.14	−12.90	−0.61	0.088
β_{13}	Group × Gaze	−3.04	4.35	−12.12	4.74	0.104
Moved						
β_{20}		5.59	3.02	0.98	12.63	0.945
β_{21}	Group	2.55	4.27	−5.23	11.51	0.877
β_{22}	Gaze	−11.80	4.85	−22.82	−3.87	0.001
β_{23}	Group × Gaze	3.17	5.46	−7.09	14.66	0.677
Looked						
β_{30}		−3.70	5.21	−15.02	5.57	0.070
β_{31}	Group	−6.95	7.77	−22.93	7.13	0.197
β_{32}	Gaze	4.04	5.25	−5.37	15.30	0.941
β_{33}	Group × Gaze	−5.68	8.13	−22.44	9.71	0.233

Note: Prob > 0 is the posterior probability that the parameter is > 0. The categories are Spoke, Moved, Looked, and No Intervention. The No Intervention category has all parameters aliased to 0.

An important lesson is that not everything can be settled using the data at hand, and often in order to understand what is wrong with a model, the original data have to be checked, and other issues examined, including the underlying theory that went into the experiment in the first place.

Having said that, it should be noted that Paananen et al. (2020) have revised the 'leave-one-out' method.[1] In our program above, the loo variable is 'log_lik_itype', and the name of the fit (in RStudio) is 'fit_cat'. In the new method, we can write

```
loo_lik_itype2 <- loo(fit_cat, moment_match = TRUE, parameter_name =
"log_lik_itype")
```

This results in a slightly worse elpd_loo = –57 indicating that the previous estimate of the predictive fit was an overestimate, but the p_loo (effective number of parameters) is slightly reduced at 13.2 (true number is 12). However, 32 of the Pareto k values are 'good', 6 are 'ok', and none are 'bad'.

This gives us another way to identify problematic observations. 'loo_lik_itype2$diagnostics$pareto_k' will give the new Pareto k values, and the original ones are stored in 'loo_lik_itype2$pointwise[,"influence_pareto_k"]'. Plot one against the other in order to see individuals that have different Pareto k values and which are the objects of suspicion.

Binomial Logistic Model

Concepts

Above, we have considered response variables that are ordered or categorical. A particular case is when the response variable is binary (i.e. just two categories). This is a situation where the response variable consists of Bernoulli trials (see Chapter 3) – that is, each outcome can be classified as 0 or 1. If there are n individuals, for each one, there is an outcome y_i which is 0 or 1. If probability of '1' is p_i, then for the ith individual, the likelihood is

$$p_i^{y_i}(1-p_i)^{1-y_i}$$

For n individuals, the likelihood will be the product of these n expressions for $i = 1, 2, \ldots, n$.

However, we are interested in how a set of independent or explanatory variables influence the p_i. In Bernoulli logistic regression, the (inverse) link function is the form we have used above:

$$p_i = \frac{1}{1+e^{-\eta_i}} \qquad (6.13)$$

where η_i is the usual linear predictor

$$\eta_i = \sum_{s=1}^{p} \beta_s x_{si}$$

for independent or explanatory variables $x_1, \ldots, x_p$.

Recall that this is the cumulative distribution function of the logistic distribution. No matter what the value of η_i, we will always have $0 \le p_i \le 1$. Notice also a positive value of β_j means that x_j is associated with an increase in p_i.

[1]https://cran.r-project.org/web/packages/loo/vignettes/loo2-moment-matching.html

An alternative referred to as a probit model is to use the cumulative distribution function of the normal distribution:

$$p_i = \o(\eta_i) \tag{6.14}$$

Bernoulli Logit Example: Bystander Intervention

In the bystander example, the simplest response variable to consider is whether or not participants intervened at all. This is coded as the binary variable 'intervention' in the results, and in the Stan program we refer to this as 'yint'. The score is '1' for intervention and '0' no intervention.

Figure 6.8 shows the proportions of interventions by the conditions. It seems that (InGroup, NoLookAt) has the highest proportion, but comparing just the two Gaze conditions, the NoLookAt is greater than LookAt. Let's examine how this works out with the Bayesian model, which makes use of the Stan function bernoulli_logit.

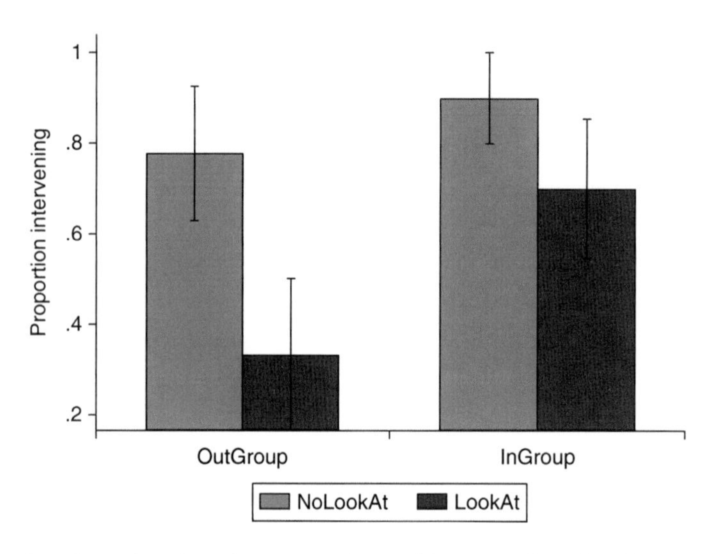

Figure 6.8 Bar chart showing the proportions of interventions by Group and Gaze. The bars are the proportions and the lines are the standard errors

Box 6.3

```
data {
    int<lower=0> n;          //number in sample
    int<lower=0> p;          //number of parameters
    vector<lower=0,upper=1>[n] g;  //Group condition
    vector<lower=0,upper=1>[n] z;  //gaze condition
    int<lower=0,upper=1> yint[n]; //binary variable for interventions
}
```

```
parameters {
  vector[p] b_int;
}

model {
  real eta;

  b_int ~ normal(0, 10);

  for(i in 1:n){
    eta = b_int[1] + b_int[2]*g[i] + b_int[3]*z[i] +
            b_int[4]*g[i]*z[i];
    yint[i] ~ bernoulli_logit(eta);
  }
}

generated quantities{
  int yint_new[n];
  real eta;
  vector[n] log_lik_yint;
  vector[n] prob;

  for(i in 1:n){
      eta = b_int[1] + b_int[2]*g[i] + b_int[3]*z[i] +
              b_int[4]*g[i]*z[i];
      yint_new[i] = bernoulli_logit_rng(eta);
      log_lik_yint[i] =  bernoulli_logit_lpmf(yint[i]|eta);
      prob[i] = 1/(1 + exp(-eta));
  }
}
```

Table 6.6 shows the results of the model. There is no interaction effect, and LookAt clearly seems to be associated with a reduction in the probability of intervention (probability = 1 – 0.021 = 0.979). There is a moderate probability (0.804) that InGroup is associated with an increase in the likelihood of intervention.

Criticizing the Model

Let's look at the results in more detail. The 'loo' results show 37 'good' Pareto k values and 1 'bad', meaning that at least one observation is particularly influential in this model. The effective number of parameters is estimated at 5.9 ± 2.1 (there are four). This is corrected somewhat by the modified 'loo', which results in 37 'good' and 1 'ok' k values with a reduced expected log pointwise predictive density (elpd) and effective number of parameters 5.5 ± 1.7.

Table 6.6 Summaries of posterior distributions of the parameters for the Bernoulli logit model for the binary intervention response

Parameter	Coefficient of	Mean	*SD*	2.5%	97.5%	Prob > 0
β_0		1.40	0.85	−0.14	3.23	0.962
β_1	Group	1.25	1.51	−1.50	4.52	0.804
β_2	Gaze	−2.17	1.13	−4.55	−0.08	0.021
β_3	Group × Gaze	0.46	1.82	−3.29	3.84	0.614

Note: Prob > 0 is the posterior probability that the parameter is > 0.

Figure 6.9 shows the posterior distributions of the probabilities (Equation 6.13) conditional on the actual observed values 'yint'. We would expect that for a good model, the main mass of the probability distribution would be near 1 for the interventions ('yint' = 1) and much lower for the non-interventions ('yint' = 0). Although the distribution for 'yint' = 1 seems reasonable, that for 'yint' = 0 is not. In fact, it is bimodal, which suggests that something fundamental is wrong with the model, such as an important missing variable.

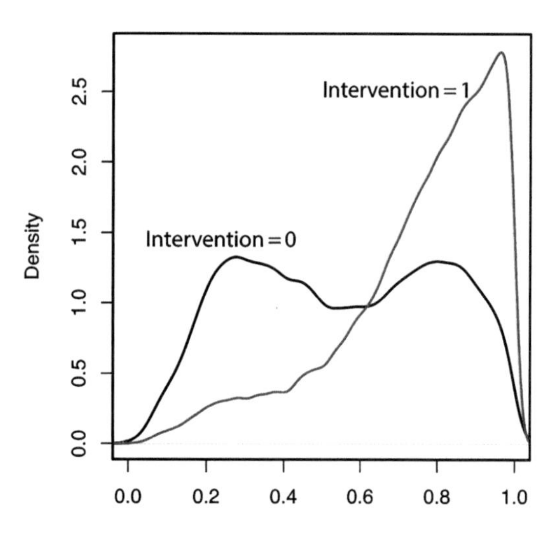

Figure 6.9 Posterior distributions of the probabilities conditional on whether the observed interventions were 0 (no intervention) or 1 (intervention)

So let's go back to the 'drawing board'. The question is whether the Gaze factor is even valid. It is supposed to represent whether the victim actually looked at the participant (bystander) during the course of the experience or not. But did the bystander even notice whether the victim was looking? If they didn't notice, then the factor is not useful.

Recall, in Table 6.1 there is a question (VictimLooked, 'vl') where participants were asked to say how much the victim was looking towards them for help. Figure 6.10 shows that there is not really much difference between the responses of those in the LookAt and those in the NoLookAt condition. The median is at the midpoint of the scale for LookAt and only one point lower than this for the NoLookAt. It seems that, after all, this factor failed experimentally; it does not by itself capture whether or not the participants were actually aware that the victim might have been looking at them.

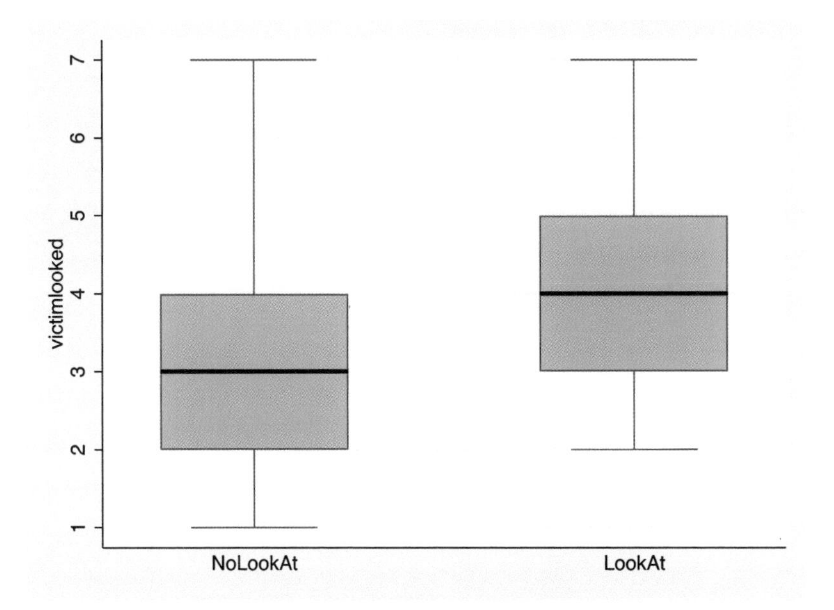

Figure 6.10 Box plot of VictimLooked by the Gaze condition

We can run the same model as above simply replacing Gaze ('z') by VictimLooked ('vl').

Box 6.4

```
data {
    int<lower=0> n;        //number in sample
    int<lower=0> p;        //number of parameters
    vector<lower=0,upper=1>[n] g;  //Group condition
    vector<lower=0,upper=1>[n] z;  //gaze condition
    vector<lower=1,upper=7>[n] vl; //VictimLooked
    int<lower=0,upper=1> yint[n]; //binary variable for interventions
}

parameters {
  vector[p] b_int;
}
```

(Continued)

```
model {
  real eta;

  b_int ~ normal(0, 10);

  for(i in 1:n){
    eta = b_int[1] + b_int[2]*g[i] + b_int[3]*vl[i] +
            b_int[4]*g[i]*vl[i];
    yint[i] ~ bernoulli_logit(eta);
  }
}

generated quantities{
  int yint_new[n];
  real eta;
  vector[n] log_lik_yint;
  vector[n] prob;

  for(i in 1:n){
      eta = b_int[1] + b_int[2]*g[i] + b_int[3]*vl[i] +
              b_int[4]*g[i]*vl[i];
      yint_new[i] = bernoulli_logit_rng(eta);
      log_lik_yint[i] =  bernoulli_logit_lpmf(yint[i]|eta);
      prob[i] = 1/(1 + exp(-eta));
  }
}
```

The results are shown in Table 6.7, which are quite different from the earlier ones in Table 6.6. In particular, there is strong evidence now that the interaction effect is associated with a higher level of intervention, Group = InGroup and participants being aware of the victim looking towards them for help (probability = 0.996) is associated with a higher probability of intervention.

Table 6.7 Summaries of posterior distributions of the parameters for the Bernoulli logit model for the binary intervention response but replacing Gaze by VictimLooked ('vl')

Parameter	Coefficient of	Mean	*SD*	2.5%	97.5%	Prob > 0
β_0		2.46	1.41	–0.03	5.50	0.974
β_1	Group	–4.27	2.43	–9.08	0.29	0.034
β_2	Gaze	–0.73	0.43	–1.67	0.03	0.030
β_3	Group × Gaze	1.64	0.71	0.36	3.12	0.996

Note: Prob > 0 is the posterior probability that the parameter is > 0.

The 'loo' analysis results in all Pareto k values as 'good', and the effective number of parameters is 4.4 ± 1.0. Although this model has a greater elpd value, the increase is small.

Figure 6.11 shows the density function for the probability conditional on the observed values. Now it looks much better, with probabilities concentrated closer to 1 for the intervention observed values, but spread out between 0 and 1 for the non-intervention values. For example, the probability of being at least 0.9 is 0.064 for the non-intervention responses and 0.305 for the intervention responses. The model is still far from perfect, since there is still bimodality for the non-intervention distribution, and a better model would be to include VictimLooked as an additional covariate rather than only replacing it with Gaze. We leave this to the reader.

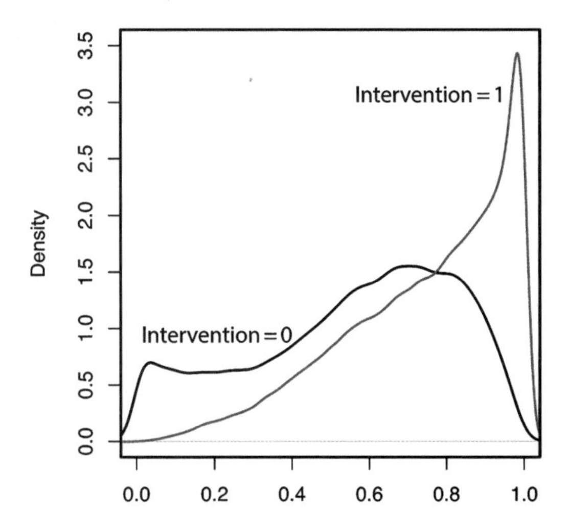

Figure 6.11 Posterior distributions of the probabilities conditional on whether the observed interventions were 0 (no intervention) or 1 (intervention) using VictimeLooked ('vl') in place of Gaze ('z')

Summary

In this chapter, we have considered typical situations where the response variable is not numerical, in the sense of not being on a ratio or interval scale. We first looked at ordinal responses, which are typical of questionnaires that employ Likert scales. We then took away even the ordering amongst the possible responses, left only with a set of unordered categories. We then essentially took away all but two categories, where the response was the observation or not of a particular outcome, thus leading to binary responses. Each type of model (they are all similar) used log odds as a basic feature, and the underlying probabilities were always constrained within the valid range of between 0 and 1.

Perhaps the most important aspect of this chapter is that we have stressed that models are produced and results obtained, but they must be criticized. The Bayesian method gives us great flexibility for this, since we can generate many different conditional posterior distributions and examine whether they meet our expectations. This is so different from standard practice that can often be seen in publications that employ the frequency-based model of statistics, where the authors carry out a significance test, report $P < 0.05$, and that's the end of it. In the Bayesian method, the assumptions are 'upfront'. What is the likelihood? This is usually derived from theoretical considerations. There is no one-size-fits-all of assuming that everything is based on a normal distribution. What are the prior distributions? A good practice, which we have not emphasized here, is to run the model several times, each time with a different (reasonable) prior to see how sensitive the results are to the prior. Sometimes the prior will be clear from the context – for example, based on past data, but often, this will not be the case, other than the analyst knowing, for example, a range of reasonable values for the prior.

When the response variable is ordinal, there is a great temptation just to forget this and treat it as if on an interval or ratio scale. This is the default approach to analysis of questionnaire data. Apart from the fact that underlying normality assumptions are unlikely to be met, this method implicitly assumes that a score of '2' is somehow double that of '1', or that the difference in the underlying phenomena is the same when the score of '4' is compared with '1', and a score of '7' compared with '4'. Whether Bayesian statistics is used or not, such methods should be avoided in order to minimize the chance of non-valid results.

In the next, concluding, chapter, we will generalize a little more. Up to now, all our response variables have been univariate, but sometimes response variables are inherently multivariate. Finally, we will describe the type of the underlying simulation algorithm that makes the Bayesian analysis possible.

Online Resources

The data, R, and Stan programs that go along with this chapter are available at

www.kaggle.com/melslater/slater-bayesian-statistics-6

Seven

MULTIPLE ISSUES

Introduction

In previous chapters, we have considered various standard models for investigating relationships between a response variable and independent or explanatory variables. We considered the general linear model, which encompasses the vast majority of techniques used in statistical analysis such as multiple regression and analysis of variance. Then we looked at a generalization of this, the generalized linear model, which moved away from reliance on normally distributed data to characterize the likelihood, and the identity relationship between the mean of the response variable and the linear predictor (a combination of the independent and explanatory variables linear in the unobservable parameters). We also considered models for responses such as binary, ordinal, and categorical that are not discrete nor continuous numerical variables.

There are multiple issues to cover in this chapter. We will generalize still further compared to the previous models, where now the response variable itself can be multivariate. Second, our models so far have been so-called 'fixed effects'. The effects over the individuals in the sample are assumed fixed, and the observations from one individual to another are assumed independent. But suppose that there are two or more observations per individual at different times; then it is clear that the observations can no longer be considered as independent, since they could be clustered by the individuals. Hence, we will consider the so-called 'mixed effect' models, where these violations of the 'fixed effects' situation are taken into account.

In previous examples, we have seen that it is important to criticize our models and find out where they do not work. One aspect of this to which we will pay more attention here is to examine the convergence properties of solutions. We will then briefly describe how the Bayesian method works in terms of the numerical simulations that solve the intractable integrals that form the solutions for the posterior distributions, and then we go on to show methods that can be used to see how well these solutions have converged.

Finally, we will make some general points about the Bayesian method and why this method is advocated in this book.

Multivariate Responses Based on the Normal Distribution

Concepts

The multivariate normal distribution was introduced in Chapter 2. We saw that it is a generalization of the univariate normal distribution. For example, when the number of variables is two, then the shape of the distribution looks like a bell and each of the marginal distributions is normal.

Recall that if $\boldsymbol{y}^T = [y_1, y_2, \ldots, y_p]$ is a vector of p random variables with a multivariate normal distribution, then the probability density function (pdf) is of the form

$$f(\boldsymbol{y} \mid \mu, \Sigma) \propto \exp\left(-\frac{1}{2}(\boldsymbol{y}-\mu)^T \Sigma^{-1} (\boldsymbol{y}-\mu)\right) \tag{7.1}$$

where $E(y|\boldsymbol{\mu}, \boldsymbol{\Sigma}) = \boldsymbol{\mu}$ and $Cov(y|\boldsymbol{\mu}, \boldsymbol{\Sigma}) = \boldsymbol{\Sigma}$.

Recall that superscript T means transpose.

In Equation (7.1), parameter $\boldsymbol{\mu}$ is a $p \times 1$ vector of means, and $\boldsymbol{\Sigma}$ is the $p \times p$ variance–covariance matrix. On the main diagonal of $\boldsymbol{\Sigma}$ are the variances, and the off-diagonal terms are the covariances. Since $Cov(y_i, y_j) = Cov(y_j, y_i)$, the $\boldsymbol{\Sigma}$ matrix is symmetric (and it is also positive definite – its determinant is positive).

Now suppose in an experiment such a random vector y is the response variable on which there are n observations, and there are k explanatory or independent variables $x_1, x_2, \ldots, x_k$. So for the ith individual, the model would be

$$
\begin{aligned}
E(y_{i1}) &= \beta_{11}x_{i1} + \beta_{21}x_{i2} + \ldots + \beta_{k1}x_{ik} \\
E(y_{i2}) &= \beta_{12}x_{i1} + \beta_{22}x_{i2} + \ldots + \beta_{k2}x_{ik} \\
E(y_{i3}) &= \beta_{13}x_{i1} + \beta_{23}x_{i2} + \ldots + \beta_{k3}x_{ik} \\
&\vdots \\
E(y_{ip}) &= \beta_{1p}x_{i1} + \beta_{2p}x_{i2} + \ldots + \beta_{kp}x_{ik} \\
i &= 1, 2, \ldots, n
\end{aligned}
\tag{7.2}
$$

This can be set out succinctly in matrix form. On the left is the set of p response variables. On the right is the set of k explanatory variables. The joint distribution of the random vectors $y_1, y_2, \ldots, y_p$ is multivariate normal conditional on the mean vector given by Equation (7.2) and also conditional on the covariance matrix $\boldsymbol{\Sigma}$.

$$
E\begin{pmatrix} y_1 & y_2 & \cdots & y_p \\ \hline y_{11} & y_{12} & \cdots & y_{1p} \\ y_{21} & y_{22} & \cdots & y_{2p} \\ \vdots & \vdots & \vdots & \vdots \\ y_{n1} & y_{n2} & \cdots & y_{np} \end{pmatrix} = \begin{pmatrix} \boldsymbol{x}_1 & \boldsymbol{x}_2 & \cdots & \boldsymbol{x}_k \\ \hline x_{11} & x_{12} & \cdots & x_{1k} \\ x_{21} & x_{22} & \cdots & x_{2k} \\ \vdots & \vdots & \vdots & \vdots \\ x_{n1} & x_{n2} & \cdots & x_{nk} \end{pmatrix} \begin{pmatrix} \beta_1 & \beta_2 & \cdots & \beta_p \\ \hline \beta_{11} & \beta_{12} & \cdots & \beta_{1p} \\ \beta_{21} & \beta_{22} & \cdots & \beta_{2p} \\ \vdots & \vdots & \vdots & \vdots \\ \beta_{k1} & \beta_{k2} & \cdots & \beta_{kp} \end{pmatrix}
$$

(The top rows of these matrices indicate the meanings of the corresponding columns, and are not part of the matrices.)

This is a straightforward generalization of the general linear model (Equation 5.4).

Prior distributions have to be placed on the $\boldsymbol{\beta}_j$ parameters and also on Σ. The first is straightforward: as usual we could give each component β_{ij} a normal distribution, or if there is prior information then whatever is appropriate.

Σ is more complex. Following the recommendations in Gelman et al. (2014) as modified in the Stan reference manual, we can decompose Σ into the product of three matrices:

$$\Sigma = \Lambda P \Lambda \tag{7.3}$$

where $\boldsymbol{\Lambda}$ is a diagonal matrix, and $\boldsymbol{P}$ is a matrix of the correlation coefficients. This becomes more clear if we write it out in full:

$$\Sigma = \begin{pmatrix} \lambda_1 & 0 & \cdots & 0 \\ 0 & \lambda_2 & \cdots & 0 \\ 0 & 0 & \cdots & 0 \\ \vdots & \vdots & \vdots & \vdots \\ 0 & 0 & \cdots & \lambda_p \end{pmatrix} \begin{pmatrix} 1 & \rho_{12} & \cdots & \rho_{1p} \\ \rho_{21} & 1 & \cdots & \rho_{2p} \\ \rho_{31} & \rho_{32} & \cdots & \rho_{3p} \\ \vdots & \vdots & \vdots & \vdots \\ \rho_{p2} & \rho_{p2} & \cdots & 1 \end{pmatrix} \begin{pmatrix} \lambda_1 & 0 & \cdots & 0 \\ 0 & \lambda_2 & \cdots & 0 \\ 0 & 0 & \cdots & 0 \\ \vdots & \vdots & \vdots & \vdots \\ 0 & 0 & \cdots & \lambda_p \end{pmatrix}$$

The first and the last matrices are the diagonal ones with 0 everywhere except on the main diagonal, and the middle matrix is a correlation matrix, with 1s on the main diagonal and the correlation ρ_{ij} on the off-diagonal (recalling that $\rho_{ij} = \rho_{ji}$ and $\rho_{ii} = 1$).

Now if we multiply out the matrices, the (i, j)th term is

$$\lambda_i \rho_{ij} \lambda_j$$

But the (i, j)th element of $\boldsymbol{\Sigma}$ is the covariance Σ_{ij}. Therefore,

$$\Sigma_{ij} = \lambda_i \rho_{ij} \lambda_j$$

Moreover,

$$\Sigma_{ii} = \lambda_i^2$$

Since the main diagonal of $\boldsymbol{\Sigma}$ contains the variances, it follows that the λ_i are the standard deviations. This now turns a difficult problem into an easier one, since now we only have to give priors for the standard deviations and the correlations, and then the prior for $\boldsymbol{\Sigma}$ is given by the product (Equation 7.3).

For the standard deviations λ_i, we can use, for example, the Cauchy distribution restricted to positive values, or the Gamma distribution, as before. For the correlation matrix, we can use the LKJ distribution introduced in Chapter 2. In Stan, in order to compute Equation (7.3), we can use the function

```
quad_form_diag(P, lambda)
```

where 'P' is the correlation matrix, and 'lambda' is the vector consisting of the λ_i.

Application

This is concerned with a study of racial bias of White people against Black (Hasler et al., 2017). The study used the technique of embodiment in virtual reality (VR), where a person in VR can have a different body to their own (e.g. when they look down towards themselves or look in a virtual mirror, they would see a virtual body replacing their own and that moves with their movements). The experimental study required participants to have a VR experience on two separate occasions, 1 week apart. On the first occasion, they would be embodied in a body that was light-skinned or dark-skinned. During the session, they would interact with another virtual human, controlled entirely by a computer program, carrying out a simple task that involved describing a set of large photographs on a wall. The character they interacted with would be either light-skinned or dark-skinned. On the second session, they would have the same experience, except that the character with whom they interacted would have different skin colour to the first session. For example, if in the first session the person interacted with a White character, in the second week it would be a Black character, or vice versa. The person themselves would have the same body on each session, either 'Black' or 'White'.

The experimenters were interested to know how the embodiment of the participants (Black or White) influenced their interaction with the other (virtual) character. In particular, the experiment relied on the finding from social psychology that when people are in harmony with each other, they tend to non-consciously mimic each other's behaviour, not exactly but approximately. For example, as one leans to one side, the other may do a similar movement, or as one scratches their nose the other may rub their chin, and so on. This is known as the Chameleon Effect reported by social psychologists Chartrand and Bargh (1999).

So in the experimental study, the virtual character had been programmed to carry out a set of specific actions (e.g. like scratching its arm). After the experiment had been completed, independent observers watched videos of the movements of the virtual bodies of the real person and the virtual character. In order to avoid bias, the observers did not know which experimental condition they were watching, because the bodies were abstracted to be skeletal only. The critical response variable was the amount of mimicry exhibited by the participants. The hypothesis was that even though all participants were 'White', those embodied as Black would mimic the Black virtual character more than the White virtual character, and those embodied as White would mimic the White virtual character more than the Black character. Putting this another way, there would be more mimicry when the participant and the virtual character both had the same virtual skin colour than when they had different skin colours.

Table 7.1 shows the first few rows of the data. The total number of participants was $n = 32$. The factor 'ownbody' is an example of a 'between-groups' factor, that is there were two different groups of individuals (all women): one group always had the Black virtual body and the other always the White virtual body. The factor 'otherbody' refers to the body of the virtual character. This was different in the two sessions. This is a within-groups factor – all participants experienced both situations – a Black (or White) partner in one session and a White (or Black) partner in the second session. The two sessions were separated by a few days.

Table 7.1 First eight rows of the mimicry data

id	ownbody	otherbody1	otherbody2	nmimicry1	nmimicry2
1	0	0	1	4	8
2	0	0	1	3	1
3	0	1	0	5	4
4	0	1	0	0	0
5	1	0	1	6	5

(Continued)

Table 7.1 (Continued)

id	ownbody	otherbody1	otherbody2	nmimicry1	nmimicry2
6	1	0	1	7	10
7	1	1	0	1	0
8	1	1	0	1	1
⋮					

Note: The factor ownbody refers to the embodiment of the participant (0 = *White*, 1 = *Black*); otherbody1 and otherbody2 refer to the skin colour of the virtual character (0 = *White*, 1 = *Black*) in the first and second sessions, respectively; nmimicry1 and nmimicry2 are the counts of the number of mimicry events in sessions 1 and 2 recorded by the observers.

Figure 7.1 shows the mimicry events. It can be seen that the mean levels of mimicry were generally higher when the own body and other body skin colours were the same, in line with the expectation based on the Chameleon theory. At first sight, it seems that participants, even though all White, were more likely to mimic their Black virtual partner when they themselves were embodied as Black than when embodied as White. This is especially noticeable in session 2.

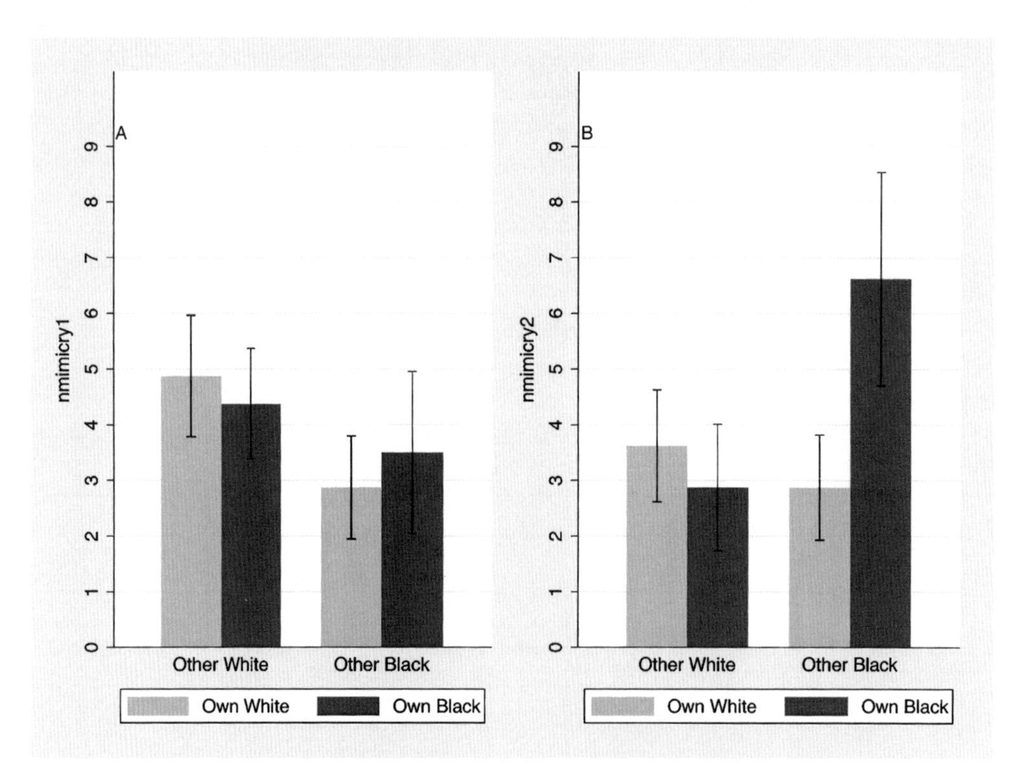

Figure 7.1 Bar charts showing the means and standard errors of the number of mimicry events for (A) the first session and (B) the second session

Recall that the response variable 'number of mimicry events' is count data and does not satisfy the requirements of normality. However, although it is not recommended (O'Hara and

Kotze, 2010) and is generally not necessary in a Bayesian context where we are far more free to choose appropriate models, just for the sake of the example we transform the count variables by a square root transformation. This has been long recommended for count data (Bartlett, 1936). This transformation makes the number of mimicry events for each separate variable compatible with normality.

Box 7.1

```
data {
    int<lower=0> n;         //number in sample
    int<lower=0> p;         //number of response variables
    int<lower=0> k;         //number of parameters
    matrix[n,k] x;          //n*k matrix
    vector[p] y[n];         //mimicry
}

parameters {
    matrix[k,p] beta;                  //the coefficients of the xs
    corr_matrix[p] P;                  //correlation matrix
    vector<lower=0>[p]  lambda;        //the standard deviations
}

transformed parameters {
  cov_matrix[p] Sigma;

  Sigma = quad_form_diag(P,lambda);  //create the covariance matrix
}

model {
    matrix[n,p] mu;

    //specify the priors for the beta
    to_vector(beta) ~ normal(0,10);

    mu = x*beta; //means, n*p matrix.

    //now for the lambda
    lambda ~ cauchy(0,5);

    P ~ lkj_corr(3.0);

    for(i in 1:n){
```

(Continued)

```
            y[i] ~ multi_normal(row(mu,i),Sigma);
        }
}

generated quantities{
    vector[p] y_new[n];
    matrix[n,p] mu;

    mu = x*beta;
    for(i in 1:n){
        y_new[i] = multi_normal_rng(row(mu,i),Sigma);
    }
}
```

The model is the multivariate normal model as set out in the previous section where the response variable observations are $[y_{i1}, y_{i2}]$, $i = 1, 2, \ldots, n = 32$, with y_1, y_2 being the square roots of the number of mimicry events, referred to as snmimicry1 and snmimicry2. The factor variables are ownbody and otherbody1 which is otherbody at the first session (at the second session, it is the opposite of what it was at the first session). In each case, White is coded as 0, and Black as 1. Hence, the interaction term ownbody × otherbody is 1 when both ownbody and otherbody are Black in the case of snmimicry1, and when ownbody is Black and otherbody is White in the case of snmimicry2.

The Stan program is shown above. The 'transformed parameters' block computes Equation (7.3). The remainder of the program just represents the model described in the previous section.

The summary of the posterior distributions of the parameters is shown in Table 7.2. As a check, the means of the distributions of the standard deviations are 0.93 and 0.99, respectively. The observed values from the data are 0.91 and 1.02, respectively. The mean of the distribution of the correlation is 0.67, and the actual correlation between the observed variables is 0.69. Hence, these particular results from the model fit well with the observed data.

Table 7.2 Summary of posterior distributions for the multivariate normal model for number of mimicry events over two sessions

Stan name	Parameter	Coefficient of	Mean	*SD*	2.5%	97.5%	N_eff	P > 0
Response	$\sqrt{\text{nmimicry1}}$							
beta[1,1]	β_{01}		2.15	0.35	1.50	2.84	273	1.000
beta[2,1]	β_{11}	ownbody	−0.26	0.48	−1.17	0.69	306	0.295
beta[3,1]	β_{21}	otherbody1	−0.75	0.49	−1.67	0.19	232	0.060
beta[4,1]	β_{31}	own × other1	0.47	0.67	−0.83	1.64	302	0.756

Stan name	Parameter	Coefficient of	Mean	*SD*	2.5%	97.5%	N_eff	*P* > 0
Response	$\sqrt{\text{nmimicry2}}$							
beta[1,2]	β_{02}		1.49	0.38	0.75	2.25	288	1.000
beta[2,2]	β_{12}	ownbody	0.91	0.54	−0.17	2.00	294	0.949
beta[3,2]	β_{22}	otherbody1	0.11	0.53	−0.88	1.23	278	0.591
beta[4,2]	β_{33}	own × other1	−1.09	0.74	−2.54	0.32	312	0.071
lambda[1]	λ_1		0.93	0.12	0.71	1.19	649	
lambda[2]	λ_2		0.99	0.13	0.77	1.27	656	
P[1,2]	P_{12}		0.67	0.10	0.45	0.84	675	

Note: The factors are ownbody (White 0, Black 1) and otherbody at session 1 (White 0, Black 1). The table shows the names of the parameters in the Stan output, the corresponding model terms, the means, standard deviations, and 95% credible intervals of the posterior distributions. N_eff is the effective sample size, and $P > 0$ is the posterior probability that the parameter is positive. All Rhat = 1.

The most important terms are the interactions. For snmimicry1, the interaction term represents the situation where the ownbody and otherbody are Black. There is moderate evidence that this is positive (probability = 0.756). There is stronger evidence that the main effect for otherbody is negative (probability = 1 – 0.06 = 0.94). This reflects what can be seen in Figure 7.1B. For snmimicry2, the situation is clearer. The probability of the interaction term being negative is 1 – 0.071 = 0.929. Here, if otherbody1 = Black, then at the second session, the other body was White. So the meaning of this is that when the ownbody was Black and the other body was White, then there was a reduction in the number of mimicry events. Here, we remind the reader that we should also be considering the effect sizes, not just the probabilities. The effect sizes can be taken as the means of the posterior distributions of the parameters. In the case of the second session, the mean is –1.09. This implies that, other things being equal, the interaction condition accounts for, on the average, that amount of reduction in the number of mimicry events. But remember, we are working on a square root scale, which complicates the interpretation. This is why many authors prefer not to transform response variables in the way we have done but, instead, choose better models, because the interpretation of the parameter estimates becomes problematic.

Before moving on to another model, we introduce some further aspects of the interpretation of the Stan output. In Table 7.2, there is the column of N_eff. This stands for 'effective sample size'. The Bayesian solutions are based on a simulation, which we will discuss later, where successive outputs of the simulation may be correlated with one another until convergence is achieved. In the case of this example, we have used a total of 2000 iterations of the simulation, but 1000 of those are warm-up (i.e. get some initial values of the simulation before actually using the results for estimation). A way to think of this is that the overall error in the simulation for each parameter is proportional to $1/\sqrt{N\_eff}$ (recall that the standard error of a sample is proportional to $1/\sqrt{N}$, where *N* is the sample size). Here, the effective sample size can be thought of as the number of uncorrelated samples drawn during the simulation. If *N_eff* is very low, then the simulation has not converged to useful values.

The header to Table 7.2 also mentions that the 'all Rhat = 1'. The simulation for solution of the integrals to find the posterior distributions is not run simply once, but several independent runs are carried out. Each run is called a 'chain'. Typically 4 chains are computed. Diagnostics for convergence are computed within each chain and then compared between chains. If the Rhat value for any parameter is much more than 1 (1.05 is the maximum recommended), then the convergence is suspect, and the result should not be used. Ideally, they should all be 1.

Figure 7.2 shows the trace plot for the fitted model. For each parameter, this shows a kind of random walk around the estimates for the parameters. If the sampling in the simulations were successful, then we should see no trends but just random fluctuations showing that the distributions have been well-sampled. Consider, for example, the distribution for the correlation P_{12}. The mean of the distribution is 0.67. We see a random walk around approximately that number. If these graphs show any clear trends, then again it is evidence that the sampling has not been successful. These graphs look reasonably 'healthy'.

Finally, we can look at the posterior distributions of the parameters. These are shown in Figure 7.3. The distributions provide more information than only looking at the results in Table 7.2. First, the distribution shapes do not demonstrate a sampling problem, and second, we can see at a glance those parameters that do not include 0 as a reasonable possibility (in the case of the distributions of the β). The 95% credible intervals also quickly show the likely ranges of the parameter values. (The R files associated with this chapter show how to obtain these results using the 'bayesplot'[1] library.)

A Mixed Model Example

Concepts

The multivariate normal model somehow doesn't capture well the actual experimental design. The participants attended on two occasions, but on each different occasion, they experienced a different set-up. The response variable was actually a count, but in order to use the multivariate normal distribution, we had to transform this by taking the square root.

Another way to model these data would be to just pool everything together and treat these data as 64 independent observations on the response variable with two factors, ownbody and otherbody (and their interaction). By taking the number of data points as 64, this reduces error by more than is warranted, since we know that each pair of observations is from the same individual and, therefore, cannot be assumed to be statistically independent. Since we could use a Poisson or negative binomial distribution for the likelihood, it overcomes the problem of response variable transformation (we can use the counts directly) but introduces another problem.

Here we consider another solution that directly models the fact that each of the $n = 32$ individuals gave two responses, thus leading to $N = 64$ observations. If we pretended that each of the N samples were independent, then the model would be of the following form:

$$y_i \sim distribution(f(\eta_i), \ldots \text{ other parameters}), i = 1, 2, \ldots, N$$

[1]https://mc-stan.org/bayesplot/

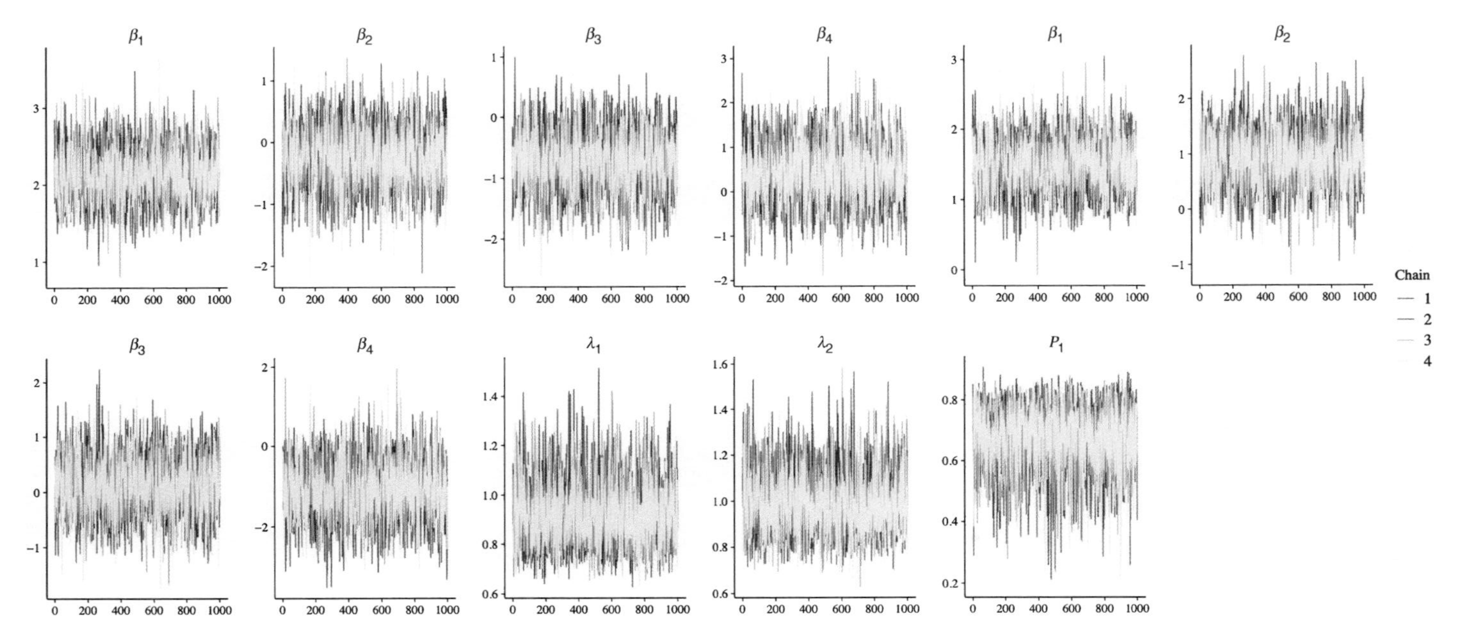

Figure 7.2 The trace plot for the multivariate normal example

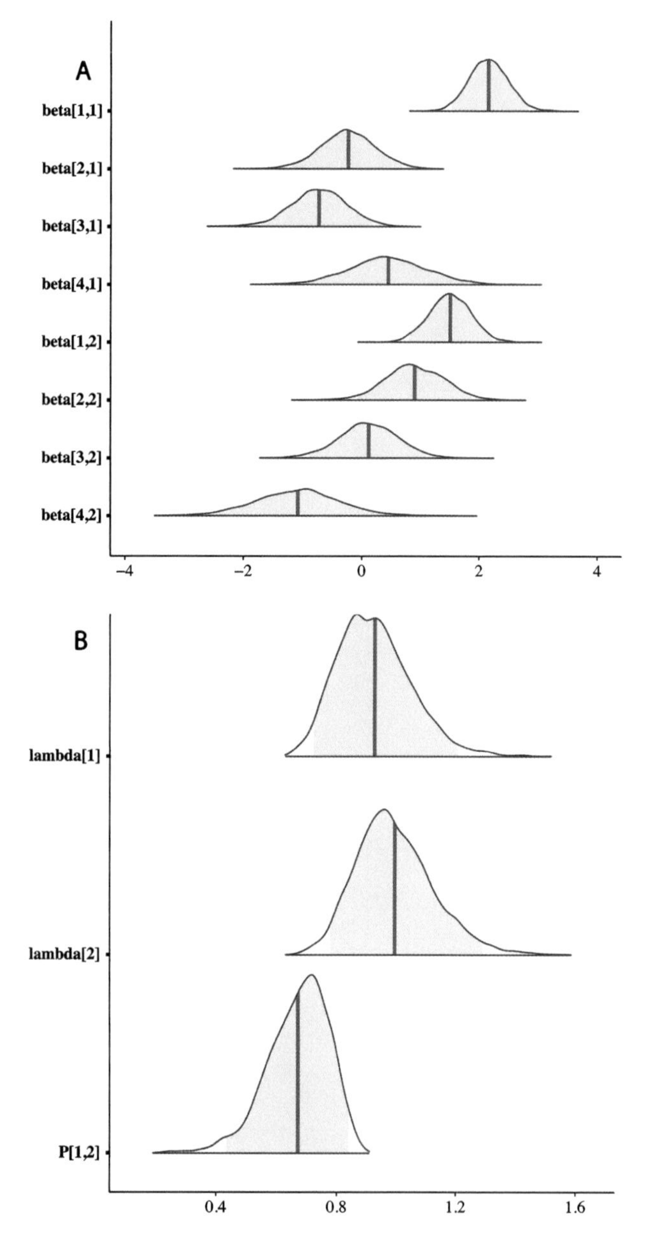

Figure 7.3 Posterior distributions of the parameters: (A) the distributions of the β and (B) the distributions of the λ and P

where η_i is the linear predictor, f is the inverse link function (e.g. exponential in the case of the log link), and *distribution* is whatever the likelihood is specified to be (e.g. in this case Poisson, or negative binomial). This is what we mean by 'pooling' all the data together, ignoring the fact that the results are not independent across the individuals.

Instead, we can introduce an additional parameter, u_i, which allows for differences between individuals. Suppose there is an array *id*, such that id_i, $i = 1, 2, \ldots, N$, is an index into the id numbers of the participants, say, $[1, 2, \ldots, n]$. For example, suppose the first four observations are 1 for individual with id 1, followed by 2 for individual with id 2. Then, $id_1 = 1$, $id_2 = 1$, $id_3 = 2$, and $id_4 = 2$. Then, instead of just the linear predictor including only the linear model in the independent variables, it would be of the form

$$u_{id_i} + \beta_1 x_{i1} + \ldots + \beta_k x_{ik}$$

$$i = 1, 2, \ldots, N$$

Here the u_i would be given prior distributions, for example *normal*$(0, \sigma_u)$, and would have posterior distributions computed from the data as usual.

In the case where the response is assumed to be normal, we have a further generalization of the general linear model (Chapter 5, Equation 5.4):

$$E(y|\beta, \boldsymbol{u}) = \boldsymbol{X\beta} + \boldsymbol{Zu}$$

where, as before, $\boldsymbol{X}$ is an $N \times k$ matrix, $\boldsymbol{\beta}$ is a $k \times 1$ vector of coefficients, $\boldsymbol{Z}$ is an $N \times n$ matrix (in this case), and $\boldsymbol{u}$ is an $n \times 1$ vector of random variables. The $\boldsymbol{\beta}$ are considered *fixed effects*, whereas the $\boldsymbol{u}$ are considered *random effects*. In Bayesian statistics, the names are slightly confusing, since all parameters have probability distributions and are therefore random variables. However, what it means is that, in this model, the $\boldsymbol{\beta}$ are unobservable but fixed and the probability distributions over the $\boldsymbol{\beta}$ represent our uncertainty about their 'true' values, whereas the $\boldsymbol{u}$ are inherently random (they are not parameters that happen to be unobservable, but actually represent a random process).

In the mimicry example that we are considering, suppose that the order of the n individuals with respect to the $\boldsymbol{u}$ vector is 1, 1, 2, 2, ..., n, n (recalling that each participant had two exposures and therefore had two mimicry values), then

$$\boldsymbol{Z} = \left.\overbrace{\begin{pmatrix} 1 & 0 & 0 & 0 & 0 \\ 1 & 0 & 0 & 0 & 0 \\ 0 & 1 & 0 & 0 & 0 \\ 0 & 1 & 0 & 0 & 0 \\ \vdots & \vdots & \vdots & \vdots & \vdots \\ 0 & 0 & 0 & 0 & 1 \\ 0 & 0 & 0 & 0 & 1 \end{pmatrix}}^{n}\right\} N$$

The variance–covariance matrix for $\boldsymbol{u}$ is assumed to be uncorrelated with that for y. This is a simple example of a 'mixed effects' model (mixed because it contains both fixed and random effects) and is a simple example of a hierarchical model. Here we only consider this case, for a further treatment see Gelman et al. (2014) and Sorensen et al. (2015).

Application to the Mimicry Data

In Table 7.1, we presented the mimicry data in 'wide form'. This means that each row corresponds to one individual, and the variables on which there was repeated data are repeated as two variables, one for the first exposure and the other for the second exposure (e.g. nmimicry1, nmimicry2). Instead, we present the data in 'long form' in Table 7.3, which is appropriate for the mixed model approach.

Table 7.3 First 10 rows of the mimicry data in long form

id	trial	nmimicry	ownbody	otherbody
1	1	4	0	0
1	2	8	0	1
2	1	3	0	0
2	2	1	0	1
3	1	5	0	1
3	2	4	0	0
4	1	0	0	1
4	2	0	0	0
5	1	6	1	0
5	2	5	1	1
⋮				

Note: Id refers to the id of the participant, trial refers to whether the row refers to their first or second exposure. The factor ownbody refers to the embodiment of the participant (0 = *White*, 1 = *Black*); otherbody refers to the skin colour of the virtual character (0 = *White*, 1 = *Black*); and nmimicry refers to the counts of the number of mimicry events recorded by the observers.

Since the mimicry response is a count variable, we could consider a Poisson or negative binomial fit, as in Chapter 5. It is left for the reader to try the Poisson; here we use the negative binomial. The Stan program has three parameter sets: beta, u, and phi (the scale parameter for the negative binomial). The link function is log, and therefore the inverse link function, which maps the linear predictor plus the u parameters into the mean, is exponential. The Stan program was run with 4000 iterations on four chains.

Box 7.2

```
data {
    int<lower=1> n;                 //number of individuals
    int<lower=1> N;                 //number of data points
    int<lower=1,upper=n> id[N];
    int <lower=1> k;                //number of parameters
    matrix[N,k] x;
    int<lower=0> y[N];              //mimicry
}

parameters {
  vector[k] beta;
  vector[n] u;
  real<lower=0> phi;
}

model {
  vector[N] mu;

  beta ~ normal(0,5);
  u ~ normal(0,5);

  phi ~ gamma(2,0.1);

  mu = x*beta;

  for(i in 1:N){
    y[i] ~ neg_binomial_2(exp(u[id[i]] + mu[i]),phi);
  }
}
```

The model corresponds to

$$\log\left(E\left(y_i|\beta,\phi,\boldsymbol{u}\right)\right) = u_{d_i} + \beta_0 + \beta_1 ownbody_i + \beta_2 otherbody_i + \beta_3 ownbody_i \times otherbody_i \tag{7.4}$$

where d_i is the id of the individual corresponding to the *i*th observation (the ith row of Table 7.3). The results are shown in Table 7.4A. The interaction term corresponding to ownbody × Black is positive (probability = 0.996), indicating that this condition leads to greatest mimicry, as can be seen from Figure 7.1. Generally in mixed models, we are not particularly interested in the random effect part, it is there to improve the model by taking into account a critical source of variation in the data. We show the summaries of the first few u values. It can be seen that some of them may be influential, for example u_4.

Table 7.4 Summary of posterior distributions for the mixed negative binomial model for number of mimicry events

Parameter	Coefficient of	Mean	SD	2.5%	97.5%	N_eff	P > 0
(A) No pooling							
β_0		0.60	1.24	–1.81	3.07	775	0.691
β_1	ownbody	0.07	1.69	–3.33	3.41	841	0.523
β_2	otherbody	–0.41	0.21	–0.82	–0.01	3061	0.022
β_3	own × other	0.73	0.28	0.19	1.30	2918	0.996
ϕ		27.38	14.86	8.18	65.20	3635	
u_1		1.36	1.28	–1.15	3.84	813	0.859
u_2		0.15	1.34	–2.56	2.78	923	0.547
u_3		1.04	1.29	–1.50	3.52	827	0.797
u_4		–5.10	3.09	–12.15	–0.15	2267	0.020
u_5		0.82	1.25	–1.67	3.24	859	0.748
u_6		1.27	1.24	–1.17	3.66	850	0.847
u_7		–1.95	1.62	–5.32	0.97	1319	0.105
u_8		–1.10	1.41	–3.91	1.62	1056	0.216
⋮							
(B) Partial pooling							
β_0		1.07	1.90	–2.62	4.99	597	0.712
β_1	ownbody	–0.24	0.38	–1.01	0.51	3741	0.267
β_2	otherbody	–0.41	0.22	–0.84	0.00	3501	0.025
β_3	own × other	0.74	0.30	0.17	1.31	3459	0.994
ϕ		27.86	15.03	7.99	64.76	3252	
μ		0.15	1.88	–3.79	3.75	593	0.149
σ		0.86	0.18	0.58	1.27	3243	
u_1		0.80	1.92	–3.10	4.45	604	0.871
u_2		–0.16	1.94	–4.14	3.59	616	0.541
u_3		0.52	1.93	–3.38	4.25	616	0.809
u_4		–1.13	2.00	–5.18	2.65	675	0.019
u_5		0.59	1.93	–3.36	4.26	615	0.735
u_6		1.01	1.92	–2.92	4.72	607	0.834

Parameter	Coefficient of	Mean	*SD*	2.5%	97.5%	N_eff	*P* > 0
u_7		–0.89	1.97	–4.94	2.82	649	0.108
u_8		–0.64	1.95	–4.62	3.06	624	0.225
⋮							

Note: The factors are ownbody (White 0, Black 1), other otherbody (White 0, Black 1). The table shows the names of the parameters using model terms, the means, standard deviations, and 95% credible intervals of their posterior distributions. N_eff is the effective sample size, and $P > 0$ is the posterior probability that the parameter is positive. All Rhat = 1.

Suppose we were interested in the outcome when ownbody is White and otherbody is White. From Equation (7.4), this corresponds to just the coefficient β_0. From this, we would conclude that there is no strong evidence of an effect when both skin colours are White (other things being equal). Figure 7.4 shows the trace plot for the parameters, including u_1. These show no trends.

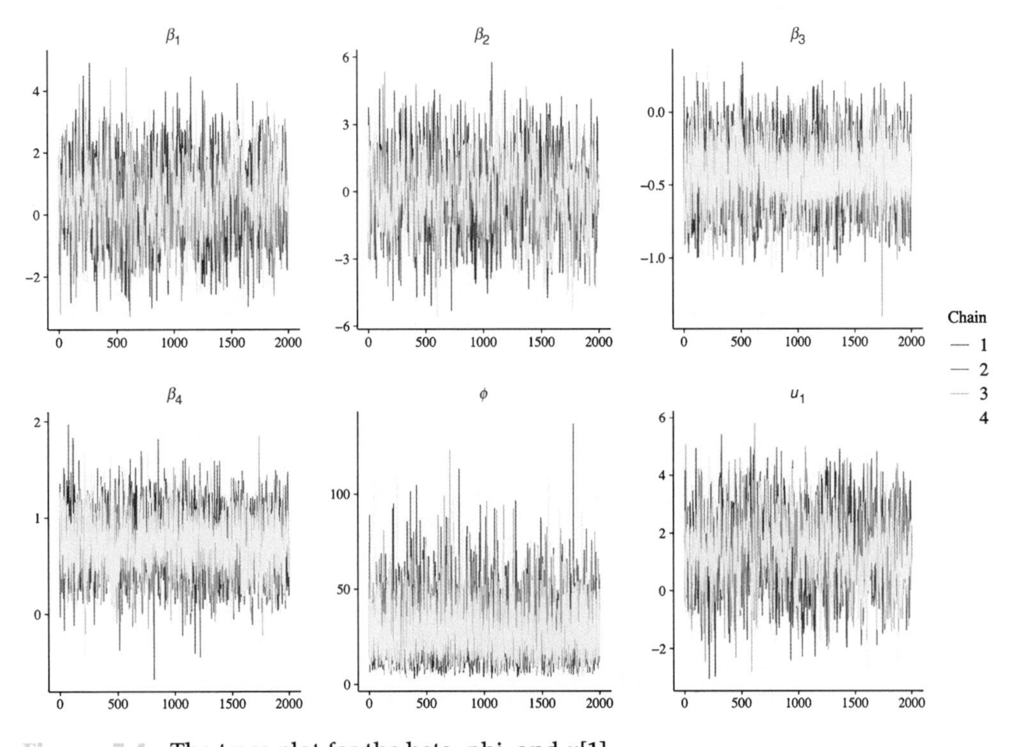

Figure 7.4 The trace plot for the beta, phi, and *u*[1]

Figure 7.5 shows the posterior distributions of the beta and phi parameters. Recall that the variance of the negative binomial distributions is $\mu + \frac{\mu^2}{\phi}$. Hence, if ϕ is large relative to μ^2, then the mean and variance of the distribution would be similar, indicating a Poisson distribution. In this case, considering Table 7.5, the observed means and variances are clearly not equal. The mean of the distribution of ϕ is 27. The range of possibilities includes also much smaller or larger values. Overall, the evidence suggests that the negative binomial distribution is a safer bet than the Poisson.

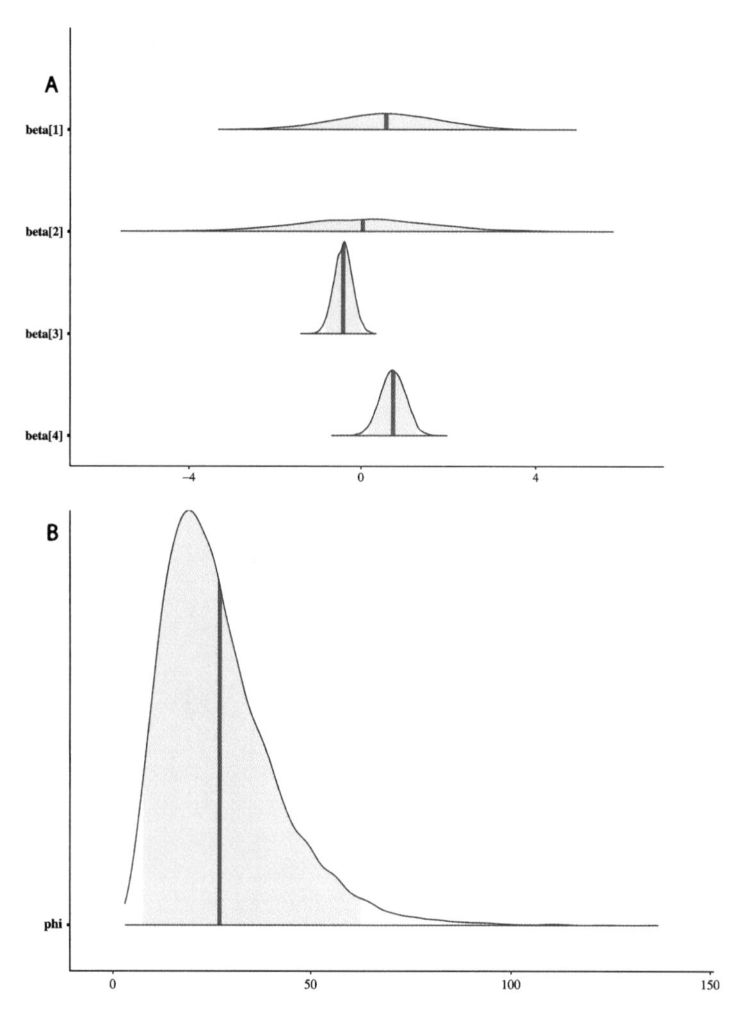

Figure 7.5 Posterior distributions of the parameters: (A) the beta (B) phi

Table 7.5 Means and variances by condition for the mimicry data

	Other White		Other Black	
	Mean	**Variance**	**Mean**	**Variance**
Own White	4.2	8.47	2.9	6.38
Own Black	3.6	8.92	5.1	23.80

Partial Pooling

The model we have used above provides an example of 'no pooling'. We have individuals and two observations on each individual. The intercept parameters are independent between the individuals. Another model that we mentioned at the start would be 'complete pooling' where we assume no differences between the individuals, so that all are pooled into one overall data set of observations. Here we introduce 'partial pooling' that does allow variation of the intercept across individuals but assumes that this arises from samples from a common distribution.

This is a particular example of what is referred to as 'hierarchical modelling'. In general, there are units of observation, and samples are taken within each unit. In our example, these units are individual people and there are two samples for each person. However, we might have multiple levels of a hierarchy, where the highest level units might be, say, countries, and then we sample between different regions in each country, and then within each region we sample different individuals, and then multiple observations for each individual. So in that case, we would have a hierarchy from country to region to individuals to samples. However, we will stay with the simpler example, where there is just the hierarchy from individuals to samples.

The idea of partial pooling can be illustrated in terms of Equation (7.4).

Pooling	**Model:** $\log\left(E\left(y_i \mid \beta, \phi, u\right)\right) =$	**Priors**
Complete	$\beta_0 + \beta_1 ownbody_i + \beta_2 otherbody_i + \beta_3 ownbody_i \cdot otherbody_i$	
None	$u_{d_i} + \beta_0 + \beta_1 ownbody_i + \beta_2 otherbody_i + \beta_3 ownbody_i \cdot otherbody_i$	$u_{d_i} \sim normal(0,5)$ independent
Partial	$u_{d_i} + \beta_0 + \beta_1 ownbody_i + \beta_2 otherbody_i + \beta_3 ownbody_i \cdot otherbody_i$	$u_{d_i} \sim normal(\mu, \sigma)$ $\mu \sim normal(0,2)$ $\sigma \sim Gamma(2,0.1)$

In the 'complete pooling' model, all the u_i are assumed to be the same and, therefore, become subsumed under the general intercept term β_0. In 'no pooling', each individual is allowed to vary independently in the intercept u_i. In the 'partial pooling' model, the intercepts vary individually but are drawn from a common distribution. (The particular distributions used are only examples.) The 'complete pooling' is too restrictive, allowing no variation between the individuals and not taking account of the fact that pairs of observations are from the same individual – that is, the hierarchy is ignored. The 'no pooling' may go too far in the other direction, not taking into account possible similarities between individuals, and with so many free parameters possibly result in overfitting. The 'partial pooling' is a compromise between these two extremes: maintaining the possibility of differences between individuals but without enforcing independence. The new parameters μ and σ in this case are sometimes referred to as *hyperparameters* and the corresponding priors as *hyperpriors*.

The corresponding Stan program is as follows with the changes highlighted.

Box 7.3

```
data {
    int<lower=1> n;                   //number of individuals
    int<lower=1> N;                   //number of data points
    int<lower=1,upper=n> id[N];
    int <lower=1> k;                  //number of parameters
    matrix[N,k] x;
    int<lower=0> y[N];                //mimicry
}

parameters {
  vector[k] beta;
  vector[n] u;
  real<lower=0> phi;
  real muu; //mean for u
  real<lower=0> sigmau; //standard deviation for u
}

model {
  vector[N] mu;

  beta ~ normal(0,5);

  muu ~ normal(0,2);
  sigmau ~ gamma(2,0.1);
  u ~ normal(muu,sigmau);

  phi ~ gamma(2,0.1);

  mu = x*beta;

  for(i in 1:N){
    y[i] ~ neg_binomial_2(exp(u[id[i]] + mu[i]),phi);
  }
}
```

The summaries of the posterior distributions are shown in Table 7.4B. The main results are the same. However, note that there are some differences with respect to the u_i. For example, for the complete pooling case (Table 7.4A), the mean of the distribution of u_4 is –5.10, whereas under partial pooling, it is –1.13, and the corresponding 95% credible interval is much narrower. However, a 'loo' analysis shows that there is not much difference between the two models in this example.

To illustrate the idea of mixed models and partial pooling in a straightforward way, we have only considered the intercept. But, of course, the same methods can be applied to the other parameters for the main and interaction effects. For these, the current example shows complete pooling. No pooling would allow different independent parameters for each individual. Partial pooling would allow different parameters per individual but where they are drawn from common distributions.

The area of hierarchical modelling is increasingly popular and important – the Bayesian approach to this is relatively straightforward. The reader could follow up some online examples[2,3] and more in-depth reading in Gelman et al. (2014, Chapter 5).

Multivariate Responses Based on the Multinomial Distribution

Concepts

Having learnt about the study on mimicry, the Department of Virtual Social Studies at the Central University of Borgonia carried out their own study. However, they used a different strategy. First, rather than adopting a mixed design (one factor between groups and the other within groups), they adopted an entirely between-groups study. Their argument was that if participants do the same task twice, then there is a chance that they understand at least partially the goal of the experimental study (because they have seen both aspects of one condition), and that this could bias the results. (The designers of the original study could argue that this is impossible since the participants could not know that their mimicry behaviour was being measured.) So in their replication experiment, there were two factors, own body and other body, as before, and 20 people were randomly allocated to each of the four cells of this design.

The second major difference was that the VR scenario was designed to specifically include 10 events where the virtual character would carry out a gesture (e.g. touching the face), and independent coders had to categorize the response of the participant to each event as: Definitely No Mimicry, Partial Mimicry, Definitely Mimicry, Don't Know. The coders were trained beforehand on the meaning of these categories, they each made their independent assessments, and then finally decided on a common answer where there was a discrepancy, after viewing the video evidence. Hence, for each of the $n = 80$ individuals in the sample, the response variable was of the form $\boldsymbol{r}_i = (r_{i1}, r_{i2}, r_{i3}, r_{i4})$, where i refers to the ith individual. The numbers $r_{i1}, r_{i2}, r_{i3}, r_{i4}$ refer to the number of assignments to Definitely No Mimicry, Partial Mimicry, Definitely Mimicry, and Don't Know, respectively. It follows that $r_{i1} + r_{i2} + r_{i3} + r_{i4} = N$ for all i, where $N = 10$ (since there were 10 events to be classified).

From Chapter 2, this matches the description of the multinomial distribution: there are 10 trials for each individual, and each trial results in an assignment to one of the four categories. Hence, the obvious likelihood function to use is the multinomial distribution:

[2] https://mc-stan.org/users/documentation/case-studies/pool-binary-trials.html

[3] https://widdowquinn.github.io/Teaching-Stan-Hierarchical-Modelling/07-partial_pooling_intro.html

$$\boldsymbol{r}_i \sim multinomial(\boldsymbol{p}_i)$$
$$i = 1, 2, \ldots, n \qquad (7.5)$$

But what are the $\boldsymbol{p}_i$? These are the probabilities of the gestures of the ith individual being classified as No Mimicry, Partial Mimicry, Definitely Mimicry, Don't Know. Hence,

$$\boldsymbol{p}_i = (p_{i1}, p_{i2}, \ldots, p_{is})$$

where, for example, p_{i3} is the probability of the gestures of the ith individual being classified as Definitely Mimicry, and in the case of our example, $s = 4$.

The $\boldsymbol{p}_i$ are unobservable parameters. An appropriate prior distribution would be the Dirichlet (Chapter 2, Section 'The Dirichlet Distribution'). This is a distribution over a *simplex*: that is, a vector where each element is non-negative and the elements sum to 1. Of course, this is the case here, since one and only one of the four categories must always be chosen so that $\sum_{j=1}^{s} p_{ij} = 1$, where $s = 4$, the number of categories. Hence,

$$\boldsymbol{p}_i \sim Dirichlet(\boldsymbol{\alpha}_i)$$
$$i = 1, 2, \ldots, n \qquad (7.6)$$

where $\alpha_i = (\alpha_{i1}, \alpha_{i2}, \ldots, \alpha_{is})$, $s = 4$ and $\alpha_{ij} > 0$.

The $\boldsymbol{\alpha}_i$ are also unobservable parameters. However, we can relate them to the question of interest: are classifications of the gestures likely to be influenced by the ownbody skin colour and the otherbody skin colour? In other words, is the probability of mimicry influenced by the factors in the experimental set-up? In general, we would have n observations on k independent or explanatory variables $x_{i1}, x_{i2}, \ldots, x_{ik}$, $i = 1, 2, \ldots, n$. For each category, there is a linear predictor as follows:

$$\begin{aligned}
\log(\alpha_{i1}) &= \beta_{11}x_{1i} + \beta_{21}x_{2i} + \ldots + \beta_{k1}x_{ki} \\
\log(\alpha_{i2}) &= \beta_{12}x_{1i} + \beta_{22}x_{2i} + \ldots + \beta_{k2}x_{ki} \\
&\vdots \\
\log(\alpha_{is}) &= \beta_{1s}x_{1i} + \beta_{2s}x_{2i} + \ldots + \beta_{ks}x_{ki}
\end{aligned} \qquad (7.7)$$

where, in this case, k is the number of independent variables, and s is the number of categories ($s = 4$). The logs are essential since it is required that the $\alpha_{ij} > 0$. As usual, the right-hand side can be expressed succinctly as

$$\begin{pmatrix} x_{11} & x_{21} & \cdots & x_{k1} \\ x_{12} & x_{22} & \cdots & x_{k2} \\ \vdots & \vdots & & \vdots \\ x_{1n} & x_{2n} & \cdots & x_{kn} \end{pmatrix} \begin{pmatrix} \beta_{11} & \beta_{12} & \cdots & \beta_{1s} \\ \beta_{21} & \beta_{22} & \cdots & \beta_{2s} \\ \beta_{31} & \beta_{32} & \cdots & \beta_{3s} \\ \vdots & \vdots & \vdots & \vdots \\ \beta_{k1} & \beta_{k2} & \cdots & \beta_{ks} \end{pmatrix} \qquad (7.8)$$

Consider the ith row of this matrix:

$$\left(\sum_{j=1}^{k} x_{ji}\beta_{j1}, \sum_{j=1}^{k} x_{ji}\beta_{j2}, \ldots, \sum_{j=1}^{k} x_{ji}\beta_{js}, \right) \tag{7.9}$$

This is exactly the expression on the right-hand side of Equation (7.8). This is useful for the construction of the Stan program.

The parameters of interest are the β, since the relative sizes of these will inform us about the influence of the x on the α and therefore on the probabilities of interest $\boldsymbol{p}$. Other things being equal, the greater a particular α, the greater the corresponding p. We can give normally distributed priors to the β, as usual. However, it is important that the β not be permitted to be too large, since from Equation (7.7) they are exponentiated, which could result in very large values of the corresponding α. What is important is not the absolute sizes of the β but the relationship and comparisons between them.

Application

Table 7.7 shows the first 10 rows of the mimicry data from the experiment. For example, participant with id 1 was recorded as definitely mimicking all 10 gestures of the virtual human character. Participant with id 2 was recorded as responding four times with No Mimicry, twice with Partial Mimicry, once with Definite Mimicry, and three times with 'Don't Know'. The corresponding experimental conditions are shown in the second and third columns.

Table 7.6 First 10 rows of the mimicry data

			Mimicry				
			None	Partial	Definite	DK	
id	ownbody	otherbody	r1	r2	r3	r4	N
1	1	1	0	0	10	0	10
2	1	0	4	2	1	3	10
3	1	1	1	1	7	1	10
4	1	0	1	3	6	0	10
5	1	1	0	2	7	1	10
6	1	0	2	1	6	1	10
7	1	1	0	1	8	1	10
8	1	0	1	2	6	1	10
9	1	1	0	0	10	0	10
10	1	0	5	0	4	1	10
...							

Note: DK = 'Don't Know'. Rows $r1$ to $r4$ are the frequencies of assignments of participant actions over 10 mimicry events. The experimental factor ownbody is 0 for White and 1 for Black. Similarly for otherbody. This was a between-groups experiment with $n = 80$ participants with an equal number in each cell of the 2 × 2 factorial design.

Figure 7.6 shows box plots of the results for the *r*1 to *r*4 values. We use box plots here in order to give an idea of the distributions rather than just the summary measures. It is clear that the greatest number of mimicry events (*r*3) was recorded for the condition ownbody Black and otherbody Black. However, there are also relatively high values (*r*3) for ownbody White and otherbody Black, and for otherbody White and ownbody White or Black.

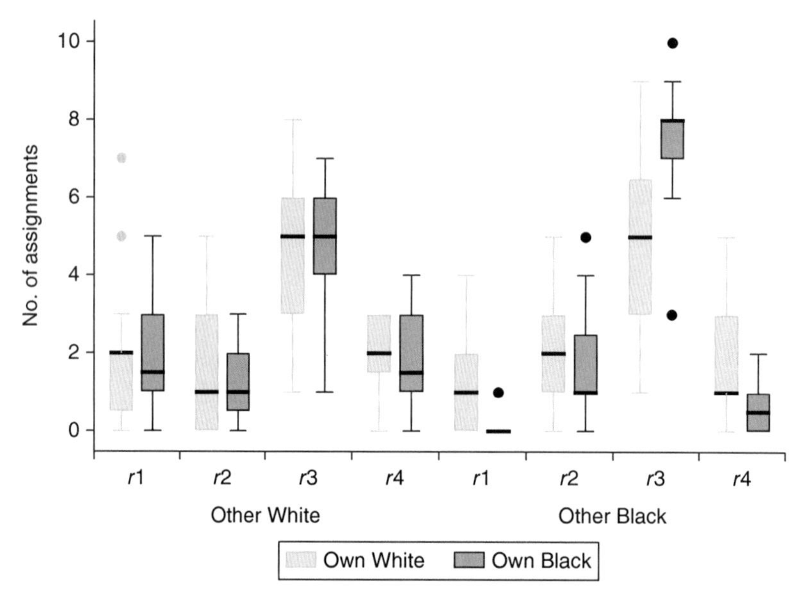

Figure 7.6 Box plots of the *r*1 to *r*4 values by ownbody and otherbody

Box 7.4

```
data {
  int<lower=0> N;              //number of multinomial trials 10
  int<lower=0> n;              //number in sample 80
  int<lower=0> k;             //number of parameters 4
  matrix[n,k] x;               //n*k matrix
  int<lower=0> s;              //length of vector of responses 4
  int r[n,s];                  //mimicry counts
}

parameters {
  matrix[k,s] b;
  simplex[s] p[n];
}
```

```
model {
  matrix[n,s] v;
  vector[s] alpha;

  to_vector(b) ~ normal(0,0.5);
  v = x*b;

  for(i in 1:n){
    alpha = exp(row(v,i))';

    p[i] ~ dirichlet(alpha);
    r[i] ~ multinomial(p[i]);
  }
}
```

In the Stan program, the 'x' matrix has each row as

[1, ownbody, otherbody, ownbody × otherbody]

for the individual. Therefore, the 'x' matrix is 80×4. The variable $s = 4$ in the data block is the number of categories. In Stan, a parameter can be declared with type 'simplex'. This means that each p is a simplex with $s = 4$ entries, and there are n of them. The model block follows Equations (7.5) to (7.9). The corresponding R commands can be found in the online code (see Online Resources around Box 7.4).

Table 7.7 shows the summaries of the posterior distributions. The interesting categories are those labelled as 'Definitely Mimicry'. There we see a substantial interaction effect (ownbody and otherbody both Black) and a substantial effect of both factors being White (b[1,3]). However, the main effects are also strongly positive. For the 'Definitely No Mimicry' category, we can see the high probability of a negative interaction effect (probability = 1 – 0.026 = 0.974), but it is also high for both being White. The distributions are shown in Figure 7.7.

Table 7.7 Summary of posterior distributions

Parameter	Coefficient of	Mean	*SD*	2.5%	97.5%	N_eff	*P* > 0
Definitely no mimicry							
b[1, 1]		0.49	0.23	0.04	0.94	1908	0.983
b[2, 1]	ownbody	0.26	0.29	−0.31	0.82	2176	0.811
b[3, 1]	otherbody	−0.17	0.31	−0.77	0.42	1855	0.282
b[4, 1]	own × other	−0.76	0.39	−1.53	0.00	1983	0.026

(Continued)

Table 7.7 (Continued)

Parameter	Coefficient of	Mean	SD	2.5%	97.5%	N_eff	P > 0
Partial mimicry							
b[1, 2]		0.33	0.24	−0.14	0.79	1965	0.915
b[2, 2]	ownbody	0.29	0.30	−0.28	0.87	2486	0.834
b[3, 2]	otherbody	0.50	0.29	−0.06	1.08	2401	0.959
b[4, 2]	own × other	0.43	0.36	−0.27	1.14	3096	0.892
Definitely mimicry							
b[1, 3]		1.29	0.23	0.85	1.72	1703	1.000
b[2, 3]	ownbody	0.54	0.27	0.01	1.07	2433	0.977
b[3, 3]	otherbody	0.48	0.28	−0.06	1.01	2526	0.958
b[4, 3]	own × other	0.79	0.34	0.14	1.46	3255	0.991
Don't know							
b[1, 4]		0.55	0.23	0.10	1.00	1780	0.992
b[2, 4]	ownbody	0.26	0.29	−0.29	0.83	2248	0.819
b[3, 4]	otherbody	0.08	0.29	−0.49	0.64	2347	0.608
b[4, 4]	own × other	−0.23	0.36	−0.94	0.48	3028	0.264

Note: The factors are ownbody (White 0, Black 1) other otherbody (White 0, Black 1). The table shows the names of the parameters using the Stan terms, the means, standard deviations, and 95% credible intervals of their posterior distributions. N_eff is the effective sample size, and $P > 0$ is the posterior probability that the parameter is positive. All Rhat = 1, and 4000 iterations were run.

Another way to look at the results is to find the probability distributions of the $\boldsymbol{p}_i$ under the restrictions that i (the individuals) are in the various conditions of the experiment.

Figure 7.8 shows the posterior distributions of the $\boldsymbol{p}_i$ under all four combinations of the factors. From this it is clear that the actions of the participants were more likely to be classified as Definitely Mimicry when both the own body and other body were Black. The other distributions are similar to one another.

The method converged well with all Rhat values being 1. The trace plots are good, and also there is a close correspondence between predicted and observed values.

The Multinomial Logit Model

Instead of indirectly modelling the relationship between the probabilities and the linear predictor through the parameters of the Dirichlet distribution, another and standard approach is to model log odds of interest. The set of categories in our example are (1) Definitely No Mimicry, (2) Partial Mimicry, (3) Definitely Mimicry, (4) Don't Know. We consider the 'Don't Know' response (4) as a baseline. The idea of the multinomial logit model is that the log odds of each condition against the baseline are equal to the linear predictors (η) involving the independent variables:

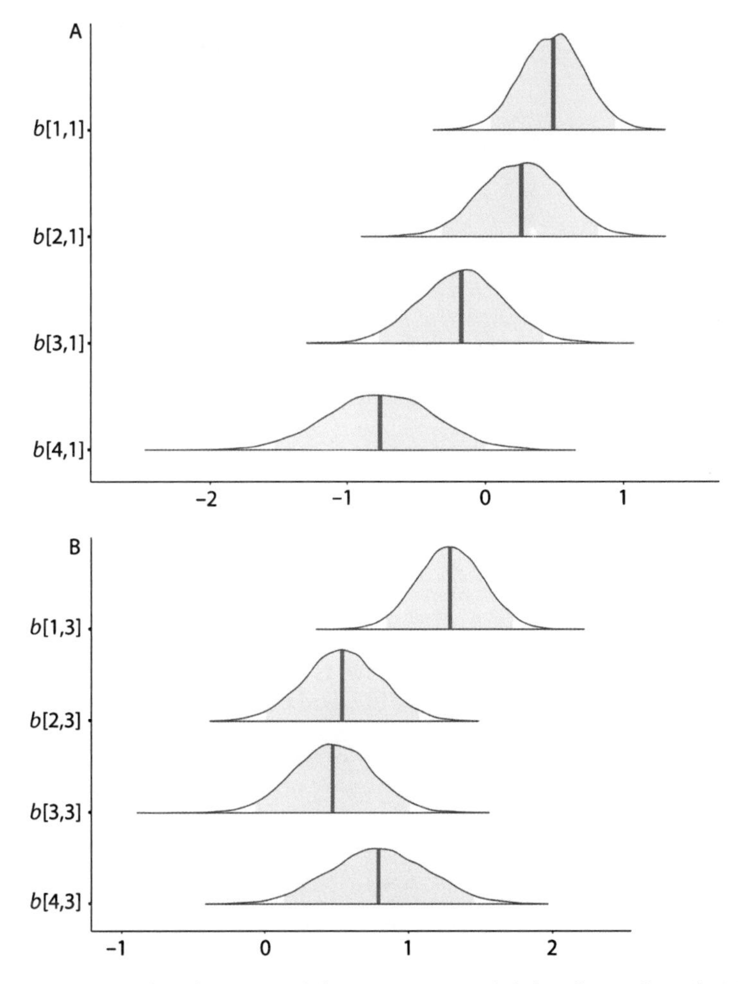

Figure 7.7 Posterior distributions of the parameters: (A) for the Definitely No Mimicry case and (B) for the Definitely Mimicry case

$$\log\left(\frac{p_{i1}}{p_{i4}}\right) = \eta_{i1} = \beta_{11}x_{1i} + \beta_{21}x_{2i} + \ldots + \beta_{k1}x_{ki}$$

$$\log\left(\frac{p_{i2}}{p_{i4}}\right) = \eta_{i2} = \beta_{12}x_{1i} + \beta_{22}x_{2i} + \ldots + \beta_{k2}x_{ki} \tag{7.10}$$

$$\log\left(\frac{p_{i3}}{p_{i4}}\right) = \eta_{i3} = \beta_{13}x_{1i} + \beta_{23}x_{2i} + \ldots + \beta_{k3}x_{ki}$$

for $i = 1, 2, \ldots, n$.

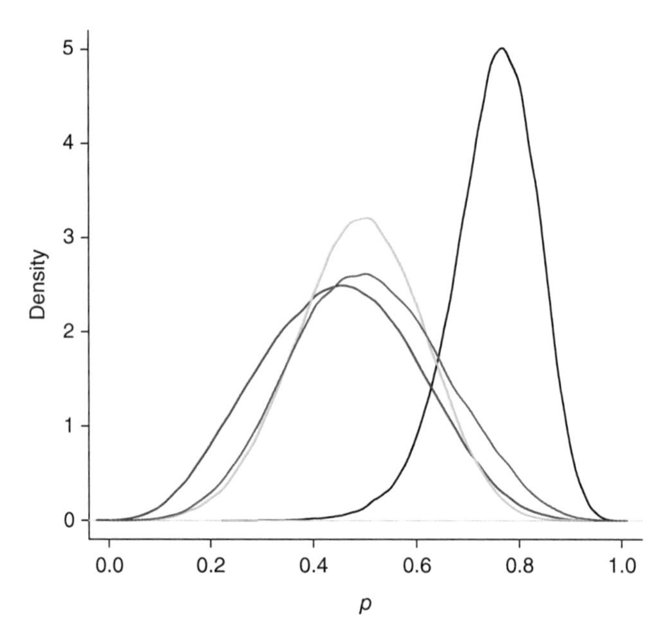

Figure 7.8 Posterior distributions for the $\boldsymbol{p}_i$ in the Definitely Mimicry case. Black is for ownbody = Black and otherbody = Black, red is for both factors set to White, green is for ownbody = Black and otherbody = White, and blue is for ownbody = White and otherbody = Black

Here, the $x_{1i}, \ldots, x_{ki}$ are the independent variables (the case of our example just consisting of the factors ownbody, otherbody, and the interaction).

Taking exponentials in Equation (7.10) and using $\sum_{k=1}^{4} p_{ik} = 1$,

$$p_{ik} = \frac{\exp(\eta_{ik})}{1 + \sum_{j=1}^{3} \exp(\eta_{ij})}, \; k = 1, 2, 3, 4 \tag{7.11}$$

This also follows from setting all the parameters of the base case (4) to 0 so that $\exp(\eta_{i4}) = 1$. It should be noticed that Equation (7.11) is the softmax function introduced in Chapter 6. It may look different, but remember that the 1 in the denominator is only because $\exp(\eta_{i4})$ has been separated out as 1.

The Stan program is a straightforward representation of this approach.

Box 7.5

```
data {
  int N;                          //multinomial total (10)
  int<lower=1> n;                 //number of individuals
  vector<lower=0,upper=1>[n] own;
  vector<lower=0,upper=1>[n] other;
  int r[n,4];
}

parameters {
  vector[4] bnon;
  vector[4] bpar;
  vector[4] bdef;
}

model {
  vector[4] eta;
  vector[4] p[n];

  bnon ~ normal(0,10);
  bpar ~ normal(0,10);
  bdef ~ normal(0,10);

  for(i in 1:n){
    eta[1] = bnon[1] + bnon[2]*own[i] + bnon[3]*other[i] +
               bnon[4]*own[i]*other[i];
    eta[2] = bpar[1] + bpar[2]*own[i] + bpar[3]*other[i] +
               bpar[4]*own[i]*other[i];
    eta[3] = bdef[1] + bdef[2]*own[i] + bdef[3]*other[i] +
               bdef[4]*own[i]*other[i];
    eta[4] = 1.0; //baseline;
    p[i,] = softmax(eta);
    r[i,] ~ multinomial(p[i,]);
  }
}

generated quantities {
  vector[4] eta;
  vector[4] p[n];
  int r_new[n,4];
```

(Continued)

```
    for(i in 1:n){
      eta[1] = bnon[1] + bnon[2]*own[i] + bnon[3]*other[i] +
               bnon[4]*own[i]*other[i];
      eta[2] = bpar[1] + bpar[2]*own[i] + bpar[3]*other[i] +
               bpar[4]*own[i]*other[i];
      eta[3] = bdef[1] + bdef[2]*own[i] + bdef[3]*other[i] +
               bdef[4]*own[i]*other[i];
      eta[4] = 1.0; //baseline;

      p[i,] = softmax(eta);

      r_new[i,] = multinomial_rng(p[i,],N);
    }
}
```

Here, 'bnon', 'bpar', and 'bdef' are the parameters for 'Definitely No Mimicry', 'Partial Mimicry', 'Definitely Mimicry', respectively, and 'Don't Know' is the baseline case.

The results are shown in Table 7.8. 'Definitely No Mimicry' was coded as the least likely compared to 'Don't Know' when both the own body and the other body were Black (probability of the corresponding parameter being negative is 1 – 0.018 = 0.982). The same interaction term is positive with high probability for being coded as 'Partial Mimicry'. Finally, the 'Definitely Mimicry', which sees the own body and other body as Black, has the highest probability (0.999). Remember that these are being compared with 'Don't Know' classifications. These results concur with the previous model that overall mimicry is more likely amongst those who experienced both their own body and their partner body as Black.

Table 7.8 Summary of posterior distributions for the multinomial logit model

Parameter	Coefficient of	Mean	*SD*	2.5%	97.5%	*P* > 0
Definitely no mimicry						
b[1, 1]		0.96	0.23	0.52	1.42	
b[2, 1]	ownbody	0.06	0.32	–0.58	0.68	0.571
b[3, 1]	otherbody	–0.13	0.34	–0.81	0.54	0.378
b[4, 1]	own × other	–1.79	0.94	–3.85	–0.09	0.018
Partial mimicry						
b[1, 2]		0.90	0.24	0.42	1.37	
b[2, 2]	ownbody	–0.25	0.35	–0.92	0.43	0.237
b[3, 2]	otherbody	0.27	0.34	–0.41	0.94	0.791
b[4, 2]	own × other	1.24	0.56	0.14	2.37	0.987
Definitely mimicry						
b[1, 3]		1.92	0.19	1.56	2.31	

Parameter	Coefficient of	Mean	SD	2.5%	97.5%	P > 0
b[2, 3]	ownbody	0.01	0.27	−0.52	0.53	0.518
b[3, 3]	otherbody	0.20	0.28	−0.36	0.74	0.773
b[4, 3]	own × other	1.54	0.47	0.66	2.50	0.999

Note: The factors are ownbody (White 0, Black 1) other otherbody (White 0, Black 1). The table shows the means, standard deviations, and 95% credible intervals of their posterior distributions. $P > 0$ is the posterior probability that the parameter is positive. All Rhat = 1, and 4000 iterations were run.

As before (Figure 7.8), we can examine the posterior predicted distributions of the probabilities p, which are shown in Figure 7.9. We can see, following Figure 7.8, that the condition when both bodies were Black resulted in the highest probability of the 'Definitely Mimicry' case.

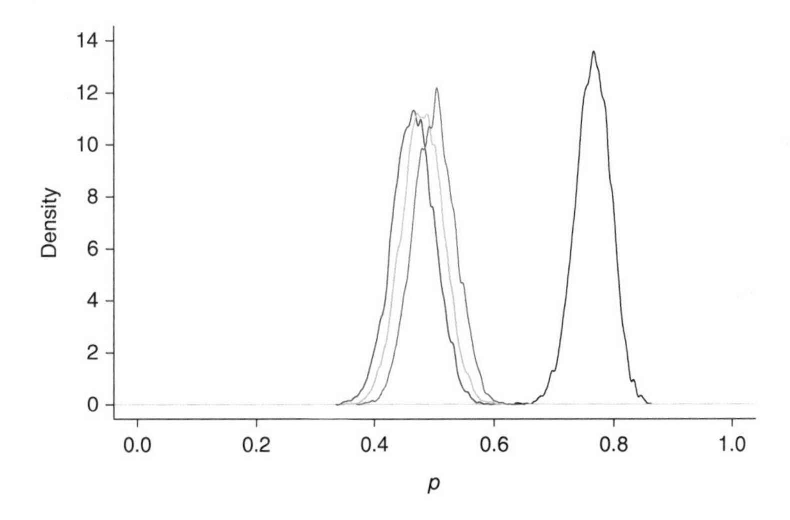

Figure 7.9 Posterior distributions for the p_i in the Definitely Mimicry case for the multinomial logit model. Black is for ownbody = Black and otherbody = Black, red is for both factors set to White, green is for ownbody = Black and otherbody = White, and blue is for ownbody = White and otherbody = Black.

The Stan system is very useful for model specification, since as we saw right from the start, the program is basically a transcription of the model into a different (but similar) notation. But how are the probabilities that are required in the model actually computed? In the next section, we take a glimpse into that.

Markov Chain Monte Carlo

Monte Carlo Concepts

Let's return to an example we considered in Chapter 3, estimating the proportion of people below the poverty line in Borgonia. Recall that we have a sample of n independent individuals, and x of them are below the poverty line. We want to find out about the proportion in the population (θ) who are below the poverty line. Here the likelihood is

$$f(x|\theta) \propto \theta^{x}(1-\theta)^{n-x},\ x = 0, 1, 2, \ldots, n$$

We use the Beta distribution as prior for θ,

$$g(\theta) \propto \theta^{\alpha-1}(1-\theta)^{\beta-1}$$

where

$$\alpha > 0, \beta > 0, \theta \in [0, 1].$$

The parameters α and β are known constants, chosen for an appropriate prior.
Therefore, from Bayes' theorem, the posterior is

$$f(\theta|x) \propto \theta^{x+\alpha-1}(1-\theta)^{n-x+\beta-1} \tag{7.12}$$

Now we know that this is also a Beta distribution, and we can therefore use this to make probability statements about θ. For example, suppose we wanted to know the probability of θ being between two bounds A and B.

Here we would require the integral

$$\int_{A}^{B} f(\theta|x)\,d\theta \tag{7.13}$$

In order to be able to do this, we would have to know the normalizing constant in Equation (7.12). We know that this is Beta($x + \alpha$, $n - x + \beta$), but supposing we did not know this. Then we would have to integrate Equation (7.12) between 0 and 1 to find that constant. Then we would have to do the integration (Equation 7.13). There is no closed algebraic form for the answer to this. So we would be forced to use a numerical method.

What we would like to do is to simulate observations on the distribution (Equation 7.12). Once we have data from this simulation, we can use the results to estimate any probability in which we are interested, or indeed find summary measures of the distribution (e.g. mean and standard deviation). In reality, we have Stan that would do all the work for us, but suppose that we did not. Here we aim to give some inkling of what is going on behind the scenes that allows Stan and other similar software to actually compute answers to these questions.

The algorithms for solving this type of problem are called Markov chain Monte Carlo (MCMC). We will explain this name as we go along. They are at the root of what makes Bayesian statistics computationally possible. Of course, in a sense they have nothing to do with Bayesian statistics, which relies on Bayes' theorem, which is a purely formal theorem that follows from the axioms of probability, but with an interpretation that is useful for statistical inference. However, MCMC methods are the machinery through which practical solutions to problems can be found. In other words, Bayesian statistics can be fully appreciated and practised with no knowledge of MCMC, but knowing something about MCMC can give the practitioner a feeling of what's going on 'under the computational hood'. It might help to understand some of the strange things we have seen before, such as trace plots.

As we have said, there are many different varieties of MCMC, and we are only going to discuss one method as an example, the Metropolis–Hasting algorithm, in fact just the Metropolis version of this. These methods go back to the 1950s (Metropolis et al., 1953) and a later generalization by Hastings (1970). For further reading, see Gelman et al. (2014, Chapter 11) for a comprehensive treatment, also Jackman (2000), Møller (2013), Van Ravenzwaaij et al. (2018), and Yildirim (2012).

So the problem is to simulate observations on a probability distribution, for example Equation (7.12). A general, though usually mathematically intractable, solution is the following. Suppose that $F(x)$ is the cumulative distribution function for a continuous random variable X. Then let $r = F(x)$. The variable r is a function of a random variable and, therefore, is a random variable itself. It turns out that the distribution r is uniform on (0,1). If F is an invertible function so that we can solve for x, then $x = F^{-1}(r)$. Hence, if we can generate observations on the uniform random variable r, then we can apply this result and obtain observations for x. Since generating pseudo-random observations on a uniform distribution is widely available in computer software, this would seem to solve the problem.

For example, suppose we wanted to generate pseudo-random observations on the exponential distribution:

$$f(x) = \lambda e^{-\lambda x},\ x \geq 0$$

The cumulative distribution function is

$$F(x) = 1 - e^{-\lambda x}$$

Setting

$$r = 1 - e^{-\lambda x}$$

and solving for x,

$$x = -\frac{1}{\lambda}\log(1-r) \tag{7.14}$$

Therefore, by generating a sequence of uniform pseudo-random numbers r_1, r_2, …, the corresponding $x_1 = -\frac{1}{\lambda}\log(1-r_1)$, $x_2 = -\frac{1}{\lambda}\log(1-r_2), \ldots$ will follow an exponential distribution. We can replace $(1 - r)$ just by r, since both r and $(1 - r)$ are uniformly distributed. As an example, the reader should generate, say, 100 pseudo-random numbers on the uniform (0,1) distribution, apply Equation (7.14), and plot the histogram. (In R, the command 'runif(n,a,b)' will generate 'n' pseudo-random observations on the uniform(a,b) distribution.)

Consider trying the same technique for the standard normal distribution. We would need to solve for x:

$$F(x)=\int_{-\infty}^{x}\frac{1}{\sqrt{2\pi}}e^{-\frac{1}{2}z^2}dz=r \tag{7.15}$$

It is not possible to find an algebraic formula, and the same will be the case for the vast majority of distributions. So we need some other method for computing such integrals.

As an aside, the reader might be wondering how values of $F(x)$ might be computed at all. Here is one simple method. Consider Figure 7.10, which shows the standard normal distribution in a bounding rectangle from –4 to 4 on the x-axis and 0 to 0.4 on the y-axis. The problem is to find $F(a)$, the area under the curve less than or equal to a; in other words, to solve the integral (Equation 7.15) for the specific value, $x = a$.

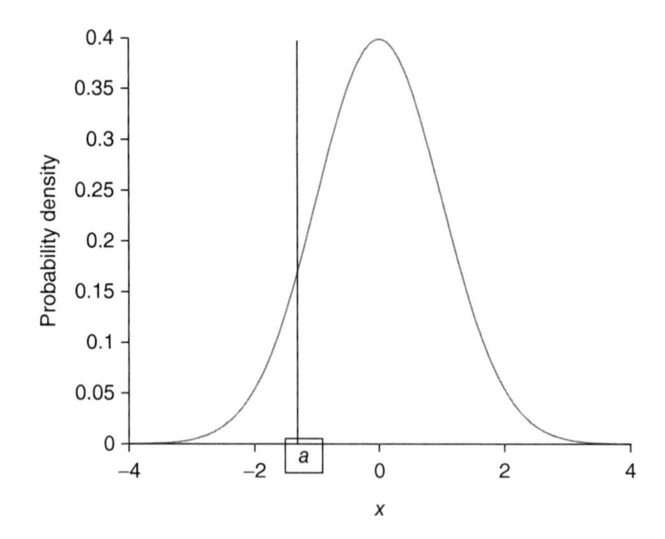

Figure 7.10 The standard normal distribution – to find the area under the curve to the left of a

The area of the bounding rectangle is 3.2. Set a counter $R = 0$. Now we generate a sequence of pairs of uniformly distributed random variables (r_{xi}, r_{yi}) and find $(8r_{xi} - 4, 0.4r_{yi})$, $i = 1, 2, \ldots, N$. These values will always be within the bounding rectangle. Let's call this pair (x_i, y_i) which will be uniformly distributed over the bounding rectangle. For each generated point, we can easily evaluate if $x_i \leq a$, and if $y_i \leq f(x_i)$, the density function under the integral sign in Equation (7.15). Each time that this is true, we increment the counter R by 1, which corresponds to the point falling under the curve and to the left of a, and otherwise we reject the point. N should be large enough that we can rely on the law of large numbers (Chapter 2) to assure that with probability 1

$$\frac{R}{N}\rightarrow p$$

where p is the probability of a random point falling under the curve to the left of a. The area that we want is therefore (approximately) $\frac{R}{N}$. We can get an idea of how 'approximate' it is, since we know from the binomial distribution that the variance of R is $Np(1-p)$, and therefore, by substituting p by $\frac{R}{N}$, we can estimate the variance and therefore the standard error.

The above procedure is a simple example of Monte Carlo integration. The term 'Monte Carlo' is used because the method apparently relies on chance, and Monte Carlo is famous for its casinos. Monte Carlo techniques solve deterministic problems by making use of the generation of pseudo-random numbers and relying on the laws of probability. Here our law was just the law of large numbers, which supports the idea that the proportion of 'hits' falling in the required area of the curve will, with probability 1, converge to the true probability of a random point falling in that area under the curve.

Markov Chain Concepts

The deterministic problem that we are interested here is to be able to solve integrals involving pdfs such as Equation (7.12) or Equation (7.15), where we cannot achieve this by deriving a mathematical formula. The approach to be taken is to generate a large number of pseudo-random observations on the distribution. Once we have those observations (think of them as forming a histogram approximating the distribution), we have a numerical solution to the problem, since suppose we want to find $P(X < a) = \int_{-\infty}^{a} f(x)dx$; then we just find the proportion of observations that satisfy $X < a$. Suppose we want the mean of the distribution; then we just find the mean of the observations, and so on.

We present the Metropolis algorithm and then explain why it works. This is based on Møller (2013, Chapter 1) and Gelman et al. (2014, Chapter 11). The problem is to draw pseudo-random observations on the distribution with pdf $f(x)$, $x \in D$ (D is a set containing the domain of values of the distribution, e.g. for the Beta distribution $D = [0, 1]$, the set of real numbers in the range 0 to 1).

Choose $p(y|x)$ as a probability distribution of y conditional on x. In other words, given a value of x, we can generate an observation y on this conditional distribution. The domain (D) of the target distribution $f(x)$ must be a subset of the domain of $p(y|x)$. A further requirement is that $p(y|x)$ must be symmetric, in the sense that $p(y|x) = p(x|y)$. For example, this could be a normal distribution with mean x:

$$p(y|x) = \frac{1}{\sigma\sqrt{2\pi}} e^{-\frac{1}{2}\left(\frac{y-x}{\sigma}\right)^2} \tag{7.16}$$

Here $p(y|x) = p(x|y)$. An even simpler example is

$$p(y|x) = 1$$

$$x, y \in [0, 1]$$

Whatever distribution is chosen, it is required to be able to draw pseudo-random observations from it. Now consider the following program that outputs a sequence of pseudo-random numbers.

Box 7.6

```
Choose an initial value x ∈ D, f(x) > 0.
Set rejected = 0
Repeat:
Generate a pseudo-random observation y from the distribution p(y|x)
Set α = min(f(y)/f(x), 1)
With probability α
        output y
        set x = y
    else
                    rejected = rejected+1
    end
end
```

The claim here is that the distribution of the sequence of values output by this procedure will eventually converge to the target distribution with pdf $f(x)$.

The algorithm starts by choosing an arbitrary initial value from the distribution. This is easy since it just needs to be a value in the domain D. The variable *rejected* is just a counter for the number of values generated that are not accepted. Then the algorithm iterates over the 'repeat'–'end' loop for as many times as specified (000s). The first step in the loop is to generate a pseudo-random value on the probability distribution that has density $p(y|x)$.

Then the value α is computed. Let's consider some possibilities. Suppose the generated value y is not in the domain D. For example, f might be a Beta distribution with $D = [0, 1]$ and $p(y|x)$ might be a normal distribution as in Equation (7.16). It is quite possible that, for example, the generated $y = -2.6$ (say). But if f is a Beta distribution, then $f(-2.6) = 0$, hence $\alpha = 0$, and this value of y would be rejected.

Now suppose that a value y is generated such that $y \in D$ and $f(y) > f(x)$. Then $\alpha = 1$, and this value of y would be accepted (i.e. output). This would be a good outcome, since according to the distribution, y has a higher likelihood than x.

Finally, suppose that $f(y) < f(x)$. Then $\alpha < 1$ and the smaller the value of α, the less likely it is that y would be output. Again, this makes sense, since small α means that y is very much less likely than x, and so there should be less chance of it being output. However, it has a non-zero chance of being output which is correct, since its *could* occur. The probability distribution may have tails with low probability, but such values are valid possibilities from the distribution.

Note that each output value y depends probabilistically only on the previous value that was output; that is, each generation of an observation from $p(y|x)$ depends on the previous x. Values output prior to the previous x have no *direct* influence. This is an example of a Markov chain.

It is a system that moves through a series of states, with transition probabilities from one state to another, where the probability of the next state only depends on the current state. The transitions here are from the distribution $p(y|x)$. Provided that each state can be eventually reached from any other state with non-zero probability, the sequence of outputs can't get stuck in a repeating cycle, and all states communicate with each other (i.e. the set of states can't be broken into subgroups); then the Markov chain will eventually settle into an equilibrium, in our case a stationary distribution. Whatever the initial value that starts off the whole process, all chains will eventually resemble one another in their statistical properties and, in particular, will have the same stationary distribution, in this case the target distribution.

To consider this issue of the stationary distribution a bit more, let X_t be the output at time t. Consider any two outputs x_a and x_b, where $f(x_b) \geq f(x_a)$. Let $g(u, v)$ be the joint distribution of (X_t, X_{t+1}). Now consider the probability that a particular transition is from x_a to x_b:

$$g(x_a, x_b) = p(x_b|x_a)f(x_a) \tag{7.17}$$

In this case, x_b is certain to be output because $f(x_b) \geq f(x_a)$.

Now consider the other way around, the probability that x_a will be output given that the previous output was x_b:

$$g(x_b, x_a) = p(x_a|x_b)f(x_b)\frac{f(x_a)}{f(x_b)} \tag{7.18}$$

Here x_a is output with probability $\frac{f(x_a)}{f(x_b)}$ since $f(x_b) \geq f(x_a)$.

However, by the symmetry of the transition density, $p(x_a|x_b) = p(x_b|x_a)$, and therefore Equations (7.17) and (7.18) are the same. This shows that

$$g(x_a, x_b) = g(x_b, x_a)$$

Therefore, transitioning from x_a to x_b or x_b to x_a has the same probability density, that is the joint distribution of any two outputs is symmetric and, therefore, have the same marginal distribution. In other words, when the stochastic process outputs observations from the target distribution, further applications of the generative algorithm make no difference to the distribution that is output, it becomes stationary (or in equilibrium).

The Metropolis algorithm is easy to implement. One of its major advantages is that because it only includes the target pdf, f, through the ratio $\frac{f(y)}{f(x)}$ it is not necessary that the normalizing constant, which ensures that the integral of the pdf is 1, be known, since the constant cancels out. Another aspect of it is that we have been thinking of this as univariate, but actually there is nothing in the algorithm or argument that requires this. The target distribution f can be multivariate, similarly the transition density would equivalently be multivariate, and nothing changes in the formulation. Hence, this method can and typically would be used to generate pseudo-random observations on multivariate distributions.

The Metropolis–Hastings algorithm, which we will not describe here, is very similar, except that it does not have the symmetry requirement on the transition density $p(y|x)$. It should be noted that Stan uses MCMC methods, known as Hamiltonian Monte Carlo and the 'no U-turn sampler' (NUTS), which are described in Chapter 15 of the Stan Reference Manual.

Metropolis Examples

Here is a simple R function that implements a version of the Metropolis algorithm. The function takes as input a starting value *x*0, the target function *f*, the transition distribution 'trans', $p(y|x)$, in a form to generate an observation *y* conditional on *x*, until *n* values have been accepted.

Box 7.7

```
metrop <- function(x0,f,trans,n) {
  #metropolis algorithm to n simulate values from the distribution
  #with pdf f, and the transition probability trans. x0 is the
  #initial value

  output <- vector()
  x <- x0
  accepted <- 0
  while(accepted < n) {
    y <- trans(x)
    alpha <- min(f(y)/f(x),1)
    if(runif(1,0,1) < alpha){
      output <- append(output,y)
      x <- y
      accepted <- accepted+1
    }
  }

  return(output)
}
```

The output of the function is the vector of accepted values. The function closely follows the description of the algorithm in the previous section.

As an example, we can use this to generate observations on the Beta distribution with parameters 2 and 8. This will require the specification of the *f* function. We do this in two parts, first a generic function that returns $x^{a-1}(1-x)^{b-1}$, where *a* and *b* are parameters $a > 0$, $b > 0$ (the function numbeta), and then a specific instance of that which returns the value for $a = 2$ and $b = 8$ ('numbeta_2_8').

Box 7.8

```
numbeta <- function(x,a,b) {
  #calculates the numerator of the beta distribution
  if(a < 0 | b < 0){
    print('parameter error in numbeta')
    return
  }
  if(x < 0 | x > 1) return(0)
  else
  return((x^(a-1))*((1-x)^(b-1)))
}

numbeta_2_8 <- function(x){
  return (numbeta(x,2,8))
}
```

Next we have to define the 'trans' function. This could be almost anything, but for $p(y|x)$, we choose a normal distribution with mean x and standard deviation 0.1. The function 'norm_x_0.1' returns a pseudo-random value on this distribution, making use of R's built-in function 'rnorm'.

Box 7.9

```
norm_x_0.1 <- function(x) {
  #returns random variate on normal(x,0.1)
  return(rnorm(1,mean = x,sd = 0.1))
}
```

Then to make use of the 'metrop' function, we can call, for example,

```
N <- 10000
x <- metrop(0.5,numbeta_2_8,norm_x_0.1,N)
```

First, we set the number of iterations to 10,000, and then call the function placing the output in the variable x. The starting value is 0.5.

Bearing in mind that the first values are dependent on the starting value, and the iteration has to settle in for a warm-up phase, we can plot the histogram of the second half of the values, against the actual pdf, as in Figure 7.11.

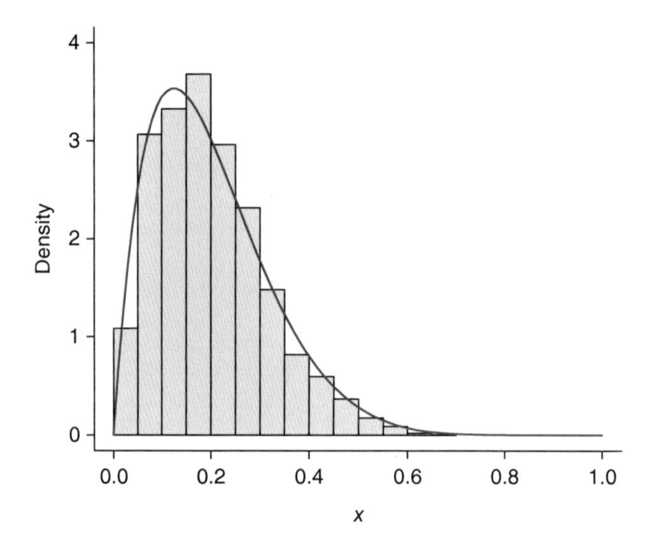

Figure 7.11 Histogram of the output from the 'metrop' function for the Beta(2,8) distribution with 10,000 iterations, using the second half of the data only, and the *normal*(x,0.1) distribution for the transition function. The red curve is the actual Beta(2,8) distribution, using the R function 'dbeta'

The result looks as if it is a good approximation to the actual pdf. The mean of the Beta(2,8) distribution is 0.2 and the standard deviation is 0.121. The mean of the x values is 0.200 and the standard deviation is 0.113.

Consider the probability of a Beta(2,8) random variable <0.2. From the distribution, this is 0.564. The proportion of x values <0.2 is 0.558. The observed distribution is reasonably accurate. The Stan program below simply simulates data on the Beta(2,8) distribution. The resulting histogram is similar to Figure 7.11. The mean of the distribution is 0.200, the standard deviation 0.121, and the probability for <0.2 is 0.562.

Box 7.10

```
//this is just to simulate data on Beta(2,8) distribution

parameters {
  real<lower=0,upper=1> theta;
}

model {
  theta ~ beta(2,8);
}
```

Our transition function, the *normal*(x,0.1) distribution, means that each successive value is going to be close to the current value. We can see the effect of this by looking at the autocorrelation function. This finds the correlations of the series of values x with itself at various lags (e.g. for a lag of d, it will find the correlations of (x_t, x_{t+d}), $t = 1, 2, \ldots$). The autocorrelation function is shown in Figure 7.12. It can be seen that not until d is more than about 20 are the correlations small. We can overcome this by *thinning* – this means that instead of including every x value in the sample, we could include only every dth value. This indeed can reduce the autocorrelation substantially. (The reader should try this.) Another way to reduce autocorrelation is to make the transition probability less dependent on x. For example, choose a much larger standard deviation, or another probability distribution for $p(y|x)$, or even a distribution that is not dependent on x at all (e.g. a uniform distribution).

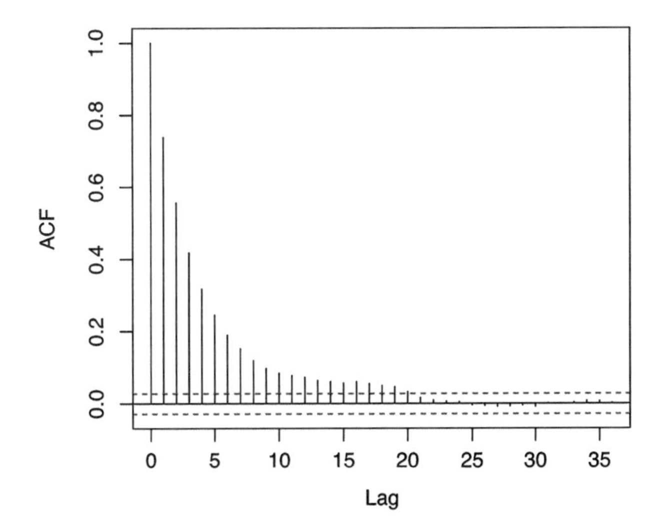

Figure 7.12 The autocorrelations for the second half of the output from the 'metrop' function for the distribution Beta(2,8)

By the way, the trace plot that we introduced above is another graphic way to consider the autocorrelations. It is just the plot of the successive values of the output. If we plot the full trace plot for the output from 'metrop' (the second half of the output), then it appears to be fine (Figure 7.13).

Plotting the autocorrelation function for the Stan output reveals no autocorrelation, recalling that it is based on the different MCMC algorithm employed by Stan.

There is no reason to just have one output from the metrop function. We could execute the function many times, each one referred to as a *chain*, each starting from a different initial value. In the code associated with this chapter, we generate m = 4 chains, each from a different random initial value. We take the second half only of each chain. Now we wish to check whether the m chains have reached the same stationary distribution (they have 'mixed'). This leads to the Rhat value mentioned earlier, which is essentially the ratio of the 'between' sum of squares

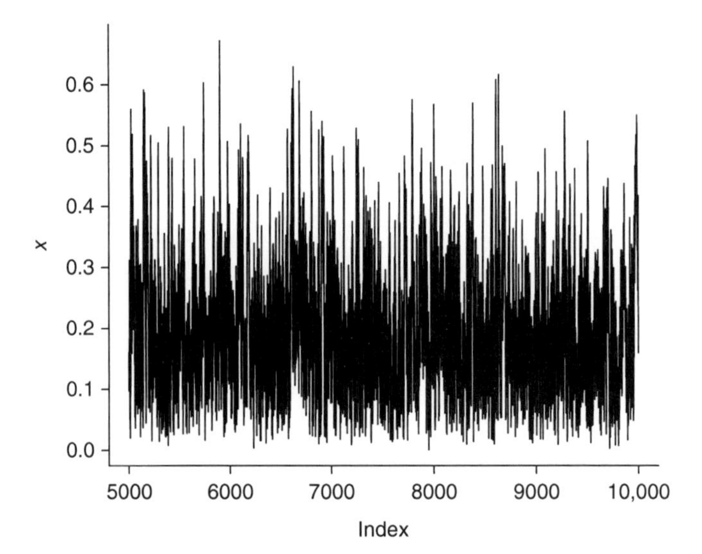

Figure 7.13 Plot of the second half of the output from the 'metrop' function

to the 'within' sum of squares of the series. This is shown in more detail in the chapter summary below – based on Gelman et al. (2014, Chapter 11).

This section has shown some of the ideas behind MCMC, the type of algorithm used by Stan and other Bayesian software. We have seen, not surprisingly, that the simple Metropolis algorithm used here does not have as good statistical properties as that used by Stan, but hopefully, it has been useful to illustrate the type of machinery that is behind these methods. As mentioned in the introduction to this section, we do not have to know about MCMC in order to carry out Bayesian analysis, but it helps.

Random Comments

A Random Walk

Let's consider an example that summarizes the message of this book. There is a particle moving in a straight line, and at each time step with probability p it moves one step to the right (+1), or one step to the left (–1) with probability $1 - p$. It starts at position $t \in [0, a]$ along the line. It will stop moving when it reaches either 0 or position $a \geq t$. We consider the probability of its reaching 0 before a, referred to as 'the probability of ruin'.

Let u_t be the probability of ruin. We can reason as follows. Ruin will occur if at the next step the particle moves to the right and is then eventually ruined, or it moves to the left and is then eventually ruined. That is,

$$P(ruin|position\ t) = P(ruin|t + 1)p + P(ruin|t - 1)\ (1 - p)$$

Therefore,

$$u_t = pu_{t+1} + qu_{t-1} \tag{7.19}$$

where $q = 1 - p$.

Equation (7.19) is a 'difference equation', and it can be solved by setting $u_t = z^t$ and solving for z:

$$z^t = pz^{t+1} + qz^{t-1}$$

Assuming $z \neq 0$, divide throughout by z^{t-1} to obtain

$$pz^2 - z + q = 0$$

Factorizing yields

$$(pz - q)(z - 1) = 0$$

Hence, there are two solutions 1 and $\frac{q}{p}$. It is easy to verify that if there are two solutions z_0 and z_1, then $Az_0 + Bz_1$ is also a solution. Hence in this case,

$$u_t = A + B\left(\frac{q}{p}\right)^t$$

We can easily find the constants A, B by using initial conditions:
Since the particle is ruined for sure at position 0,

$$u_0 = A + B = 1$$

Since there would for sure not be ruin if the particle were at position a,

$$u_a = A + B\left(\frac{q}{p}\right)^a = 0$$

Hence,

$$B = \frac{1}{1 - \left(\frac{q}{p}\right)^a}$$

and

$$A = \frac{-\left(\frac{q}{p}\right)^a}{1-\left(\frac{q}{p}\right)^a}$$

Therefore, we have the solution

$$u_t = \frac{\left(\frac{q}{p}\right)^t - \left(\frac{q}{p}\right)^a}{1-\left(\frac{q}{p}\right)^a} \tag{7.20}$$

This is a simple example of what is referred to as a 'random walk'. The MCMC method described above is a more complex random walk.

An interpretation called 'Gambler's Ruin' is where a gambler starts out with money t€ and makes a series of bets of 1€, with probability p of winning 1€ or $1 - p$ of losing 1€. The gambler decides in advance to stop playing if a€ are reached, or be forced to stop playing when 0 is reached. u_t is then the probability of being ruined (reaching 0 before a).

If $q = p = \frac{1}{2}$, then it is easy to show (for those interested consider L'Hospital's rule) that Equation (7.20) becomes

$$u_t = 1 - \frac{t}{a} \tag{7.21}$$

This is a very nice application of probability theory. We can make predictions about what is likely to happen in the world, based purely and ultimately on the axioms of probability, introduced in Chapter 1. For example, suppose our gambler started out with 10€ and decided to stop on reaching 15€, or when ruined, and each time the bet was 'evens': that is, that $p = 1/2$. Then from Equation (7.21), the probability of ruin is 1/3.

Think carefully about what this means before rushing to the casino and using this strategy. It seems to imply that the chance of winning is 2/3, which is pretty good. If you went to play once only, then this would make sense. However, if you went to the casino and employed this strategy many times, then the expected long run gain would be

$$5 \times \left(\frac{2}{3}\right) - 10 \times \left(\frac{1}{3}\right) = 0$$

All of this shows how we can make probability statements about relatively complex events by deriving the results solely from the laws of probability. However, any actual solution to a real problem relies on knowing p. If we do not know p, then we are left with Equation (7.20). Of course, we could investigate various situations asking the question: What if p were ...?

And then derive solutions based on this. However, in real situations, unless p were set by an authority (e.g. a casino in the case of Gambler's Ruin), we cannot use Equation (7.20) to make actual specific predictions about the world.

Observables and Unobservables

How do we find p? The answer is that we cannot. p is a parameter that we have referred to as an unobservable. It is not directly observable, except through observations that are dependent on p. For example, by observing a random walk we can record the number of occasions in which the particle steps to the left or right, and then use Bayes' theorem (as in Chapter 3) to make probability statements about it. Having found a posterior distribution for p, we could, in the context of Stan, also use the 'transformed parameters' block to find posteriors for Equation (7.20).

The vital insight here is that we can apply the same machinery of probability theory to the unobservable p as we can to a physically realizable process such as the position of the particle on the line in the case of a random walk. The position of the particle is a random variable with respect to physical reality. The unobservable value of p is a random variable – in our minds. But we can use the laws of probability to make inferences about it, even though we can never observe it (except at that mythical realm called 'at infinity' – which might as well be 'over the rainbow').

Another way to think about this is that the random variable representing the position of the particle on a line is observably random. One moment $t = 5$, then next it is 4, and the next it is 3, and then it is 4 again, and so on. In principle, we can compute in advance all the probabilities of these occurrences, or the occurrence of any sequence. However, although also a random variable, in the sense that any range of values of p will have a corresponding probability, p is actually considered as a fixed but unknown quantity, a parameter of the model. It is *the* probability of moving one step to the right. (Of course, p may actually vary over time, but that process would be modelled ultimately by relying on other parameters which themselves are fixed unobservables.)

This is a major difference between Bayesian and frequentist interpretations of probability as used in classical statistical inference. In the latter, p is also a fixed unknown parameter. But we are not allowed to make probability statements about it. In order to make probability statements, we would have to consider many parallel universes, where p varied between them, and thereby form a probability distribution over the range of values of p. But it would be difficult to access such data.

Confidence Intervals

The reader, probably exposed to classical, frequentist statistics may object: What about the 95% confidence interval? Isn't this a probability statement about p? Let's consider this. Suppose that there are n observations of the particle's movements, and r of those are where it moved to the right, with n 'large' (e.g. at least 30). Let $\hat{p} = \frac{r}{n}$ be the usual unbiased and consistent estimator for p, with $\hat{q} = 1 - \hat{p}$. Then the 95% confidence interval for p is

$$\hat{p} \pm 1.96\sqrt{\frac{\hat{p}\hat{q}}{n}} \tag{7.22}$$

$\hat{p}$ is a random variable. Hence, the bounds of the confidence interval are random variables. So what this interval really means is that if we repeated a large number of times the observation of n trials and each time computed these bounds, then of course each time the bound would be different, and 'in the long run' the 'true' value of p would be between the two computed limits 95% of the time. This does not lead to a probability statement about p; that is, we cannot say that the probability of p being between the two limits which we have observed in our actual *single* run is 0.95. Rather, this is a long-run probability statement about a property of the limits. This tells us nothing about the relationship between p and the particular interval that we have observed.

This is quite different from a 95% credible interval, which is an actual probability statement about p. From the prior distribution of p, we can find a 95% credible interval; that is, from the distribution of p two values L and U such that $P(L < p < U) = 0.95$. Then having observed some data from the posterior distribution of p, we can find two limits such that $P(L' < p < U'|data) = 0.95$.

In practice, a credible interval may give very similar values to a confidence interval – for example, where a uniform or improper prior has been used, and especially when n is large, but this is not the point. The point is what it means.

In case of the confidence interval, we have to postulate a large number of events that have never happened and place our findings in the context of these. We imagine 'if there had been a large number of repetitions then in 95% of those the limits would contain the true value', but in fact there are not many repetitions, there is only the one set of data. In the case of the credible interval, there is a prior credible interval and a posterior one conditional on the data. We only rely on the data that we have, not what might have been in parallel universes.

Significance Tests

Returning to Gambler's Ruin, let's suppose the game is roulette and the bet each time is on red. There are 37 possible outcomes on each spin of the wheel, and 18 of these are red, 18 black, and 1 zero (which is neither red nor black). Hence, if all is fair, then $p = 18/37$ for any bet on red or black. Since there may be doubts about 'fairness', we wish to test the hypothesis that $p = 18/37$, or in general, that $p = p_0$ (a specific value such as 18/37). We observe n (large) spins of the wheel and find r times that red occurs. In classical statistics, this would be set up as follows:

H_0: $p = p_0$.
H_1: $p \neq p_0$.

H_0 is the 'null hypothesis' – it is something specific on which we can base subsequent calculations, and typically, it is the hypothesis that we want to be able to reject. There may be different 'alternative hypotheses' (H_1); here we have chosen a generic one, which is simply the complement of the null hypothesis. There is a very elegant mathematical theory behind such tests (check out the Neyman–Pearson lemma), but it boils down to the following in this case. The observed random variable is r, and we choose a critical region R such that

$$P(r \in R|H_0) = \alpha$$

where α is fixed in advance and small, typically 0.05 or less. It is called the 'significance level'. Informally, the argument is that if we have observed such a low probability event (r being in the critical region), then either we can go along with that and accept that we have observed such an event, or we can reject the premise on which it is based, namely that H_0 is true. Hence, the usual procedure is to reject H_0. In this particular case, we can use the normal approximation to the binomial distribution, and compute

$$z = \frac{\hat{p} - p_0}{\sqrt{\frac{\hat{p}\hat{q}}{n}}} \sim normal(0,1) \tag{7.23}$$

$$P(|z| > 1.96|H_0) = 0.05 \tag{7.24}$$

where $|z|$ is the absolute value of z. Hence $|z| > 1.96$ is the critical region in this case. Notice how Equation (7.22) can be derived from Equation (7.23), and Equation (7.24) follows from the properties of the standard normal distribution. If we compute the confidence interval and reject H_0 when p_0 is not within its limits, then this would be exactly equivalent to the procedure of this test. Hence, the confidence interval and significance test are mathematically equivalent.

Let's again consider what this actually means. If we carried out a large number of repetitions of this procedure (i.e. observed n spins of the wheel and carried out the test), then if H_0 were true, 5% of the time this test would (wrongly) reject H_0. Sometimes the 'significance level' α is thought of as the probability of H_0 being true. It is not. It has nothing even remotely to do with that. It is the proportion of times in the long run that we would (wrongly) reject H_0 if in fact it were true.

Again, it should be noted that with large n and uniform or improper priors a Bayesian method would be likely to come to the same conclusion as this method, where in the Bayesian interpretation, the probability of H_0 being true would be α.

In the classical interpretation, we cannot even speak of $P(H_0)$, or if we did, there are only two possible answers $P(H_0) = 0$ or $P(H_0) = 1$ (only we don't know which). To consider $P(H_0)$, we would have to resort to the parallel universes where in some H_0 is true and in others H_0 is false.

Principles

The conclusions of so much scientific research are based on this type of procedure: null hypothesis significance tests. Their apparent simplicity belies the complex mathematical machinery and assumptions that are behind them. Reading a scientific paper, especially say in psychology, sooner or later you see something like '$t = 2.8$, $P < 0.05$'. This is so simple and standard that readers do not give it a second thought. Immediately, the reader is complicit in the implicit idea that it has been shown that the probability of the null hypothesis is <0.05.

Bayesian methods have no corresponding universal and simple manner of presentation. The concepts in themselves are much simpler to understand, without any appeal whatsoever to 'what might have happened' in the parallel universes, mysterious entities such as 'confidence intervals', which everyone knows theoretically are not probability statements but which are essentially read as such, or 'significance tests', which everyone knows are not based on the probability of the null hypothesis but which people unconsciously read as being so. However, the Bayesian method has no standard method of presentation. The classical method relies on assumptions that are rarely brought to the surface, such as reliance on normality, equal variances, and so on. In the Bayesian method, we have to explicitly state the model. It does not have to be based on a normal distribution at all. In the Bayesian method, we do not end up with a simple statement such as '$P < 0.05$', but instead we have (a set of) posterior distributions. Even in a straightforward case of inferences about a mean from a single sample, we will have three posterior distributions: the joint distribution of the mean and standard deviation and two marginal posterior distributions derived from that – one for the mean and one for the standard deviation. The method of presentation we have put forward in this book is to produce a table that summarizes the posterior distributions (mean, standard deviation, credible interval, and a probability statement about the parameter – e.g. that it is > 0) and also to display posterior distributions where this would help in understanding. There is no test of 'rejection', there is no 'null hypothesis', there are only posterior probability distributions, and it is up to the reader to interpret these in relation to the underlying questions.

Moreover, we have emphasized the importance of criticizing the model. Examine the predicted posterior distribution of the response variable, look at the impact of alternative priors, find out where and why the model does not work. As Groucho Marx said, 'Those are my principles, and if you don't like them ... well I have others.' We can say the same about prior distributions. All this requires a lot more work than simply calculating a t test value and declaring '$P < 0.05$'.

Summary

A Complement to Stan: rstanarm

Throughout this book, we have used Stan because of the direct mapping between the specification of a model and the corresponding Stan program, and therefore it is best for pedagogical reasons. Moreover, Stan straightforwardly supports models that might be considered as 'non-standard' – due to its flexibility – such as multiple response variables in the same model. There is a complementary package to Stan called rstanarm.[4] This uses Stan for computation, but its notation is quite different, and it concentrates on the standard models, essentially the general linear model (regression, analysis of variance) and the generalized linear model that we

[4] https://mc-stan.org/rstanarm/

considered in Chapter 5. The goal of rstanarm 'is to make Bayesian estimation routine for the most common regression models that applied researchers use'.[5] Once readers get into statistical analysis using Stan, for 'everyday' common models they might consider using rstanarm for its simplicity (for an introduction see also Muth et al., 2018).

Liberating

As we have observed, the Bayesian method is actually liberating. We are not stuck with normal distributions or even models that are linear in the parameters. Provided we can find a convergent solution, we are free to postulate many different types of model. In classical statistics what you have to know are lots of tests and their assumptions. In Bayesian statistics, there are no tests, there is only one principle: Bayes' theorem, the idea of updating prior probabilities from observed data. Instead of knowing tests, the practitioner has to know probability distributions. Hence the importance of Chapter 2. The more that you know and understand about probability distributions and their properties, the more likely it is that you can provide a suitable model for the data (a likelihood), and you will have a greater choice for the specification of prior distributions that reflect your uncertainty about the values of parameters.

In Borgonia, they have a saying, 'Learn distributions, not tests!' At the end, this is the most important lesson. Bayesian statistics does not rely on cookbook methods: collect data, run an analysis of variance, find where $P < 0.05$, and publish the results. Here instead, we must have a critical engagement and interaction between models and data. Hence, we must know something about distributions and their properties. A program such as Stan will do the actual computational work for us, but it is up to us to specify the model and interpret the results. We hope that this book will help towards that end in your future engagement with statistics.

The Rhat Value

Suppose the jth chain ($j = 1, 2, \ldots, m$) consists of values x_{ij}, $i = 1, 2, \ldots, n$, where n is the number generated in each chain (after removing the first half, the 'warm-up' values). Then, the between sum of squares is

$$B = \frac{n}{m-1}\sum_{j=1}^{m}\left(c_j - c\right)^2$$

where c_j is the mean of the jth chain, and c is the grand mean over all chains. There is a divisor of $(m - 1)$, since then with the sum of squares this makes the sample variance, and a multiplier of n since each of the c_j is the mean of n values. Hence, B is simply a measure of how much variation there is between chains.

[5]https://mc-stan.org/rstanarm/articles/rstanarm.html

The within sum of squares is just the mean within-chain variance:

$$W = \frac{1}{m}\sum_{j=1}^{m} s_j^2$$

where s_j^2 is just the sample variance of the *j*th chain (each using the $(n-1)$ divisor). Then a weighted sum of B and W is defined as

$$V = \left(\frac{n-1}{n}\right)W + \left(\frac{1}{n}\right)B$$

Note that

$$\frac{V}{W} = \left(\frac{n-1}{n}\right) + \left(\frac{1}{n}\right)\frac{B}{W}$$

Finally,

$$\hat{R} = \sqrt{\frac{V}{W}}$$

If the between and within sums of squares are equal, then $\hat{R} = 1$, corresponding to a situation where the variation is similar amongst the m chains. Also $\hat{R} \to 1$ as $n \to \infty$.

Online Resources

www.kaggle.com/melslater/slater-bayesian-statistics-7a
www.kaggle.com/melslater/slater-bayesian-statistics-7b

REFERENCES

Bartlett, M. (1936). The square root transformation in analysis of variance. *Supplement to the Journal of the Royal Statistical Society, 3*(1), 68–78. https://doi.org/10.2307/2983678

Chartrand, T. L., & Bargh, J. A. (1999). The Chameleon effect: The perception-behavior link and social interaction. *Journal of Personality and Social Psychology, 76*, 893–910. https://doi.org/10.1037/0022-3514.76.6.893

Chetty, R., Stepner, M., Abraham, S., Lin, S., Scuderi, B., Turner, N., Bergeron, A., & Cutler, D. (2016). The association between income and life expectancy in the United States, 2001-2014. *JAMA Journal of the American Medical Association, 315*(16), 1750–1766. https://doi.org/10.1001/jama.2016.4226

Cifarelli, D. M., & Regazzini, E. (1996). De Finetti's contribution to probability and statistics. *Statistical Science, 11*(4), 253–282. https://doi.org/10.1214/ss/1032280303

Cumming, G. (2014). The new statistics: Why and how. *Psychological Science, 25*, 7–29. https://doi.org/10.1177/0956797613504966

Darley, J. M., & Latané, B. (1968). Bystander intervention in emergencies: Diffusion of responsibility. *Journal of Personality and Social Psychology, 8*, 377–383. https://doi.org/10.1037/h0025589

Debnath, L., & Basu, K. (2015). A short history of probability theory and its applications. *International Journal of Mathematical Education in Science and Technology, 46*, 13–39. https://doi.org/10.1080/0020739X.2014.936975

Etz, A., & Wagenmakers, E.-J. (2017). JBS Haldane's contribution to the Bayes factor hypothesis test. *Statistical Science, 32*(2), 313–329. https://doi.org/10.1214/16-STS599

Feller, W. (1957). *An introduction to probability theory and its applications* (2nd ed.). Wiley.

Fishburn, P. C. (1968). Utility theory. *Management Science, 14*, 335–378. https://doi.org/10.1287/mnsc.14.5.335

Gelman, A., Carlin, J. B., Stern, H. S., Dunson, D. B., Vehtari, A., & Rubin, D. B. (2014). *Bayesian data analysis* (3rd ed.). CRC Press. https://doi.org/10.1201/b16018

Gelman, A., Hwang, J., & Vehtari, A. (2014). Understanding predictive information criteria for Bayesian models. *Statistics and Computing, 24*(6), 997–1016. https://doi.org/10.1007/s11222-013-9416-2

Hagle, T. (1995). *Basic math for social scientists: Concepts*. Sage.

Hagle, T. (1996). *Basic math for social scientists: Problems and solutions*. Sage.

Hasler, B., Spanlang, B., & Slater, M. (2017). Virtual race transformation reverses racial in-group bias. *PLOS ONE, 12*(4), e0174965. https://doi.org/10.1371/journal.pone.0174965

Hastings, W. K. (1970). Monte Carlo sampling methods using Markov chains and their applications. *Biometrika, 57*(1), 97–109. https://doi.org/10.1093/biomet/57.1.97

Jackman, S. (2000). Estimation and inference via Bayesian simulation: An introduction to Markov chain Monte Carlo. *American Journal of Political Science, 44*(2), 375–404. https://doi.org/10.2307/2669318

Jeffreys, H. (1998). *The theory of probability*. Oxford University Press.

Kahneman, D. (2011). *Thinking, fast and slow*. Macmillan.

Kropko, J. (2015). *Mathematics for social scientists*. Sage.

Latané, B., & Darley, J. M. (1968). Group inhibition of bystander intervention in emergencies. *Journal of Personality and Social Psychology, 10*(3), 215–221. https://doi.org/10.1037/h0026570

Levine, M., & Manning, R. (2013). Social identity, group processes, and helping in emergencies. *European Review of Social Psychology, 24*(1), 225–251. https://doi.org/10.1080/10463283.2014.892318

Lewandowski, D., Kurowicka, D., & Joe, H. (2009). Generating random correlation matrices based on vines and extended onion method. *Journal of Multivariate Analysis, 100*(9), 1989–2001.

Little, R. J., & Rubin, D. B. (2019). *Statistical analysis with missing data* (Vol. 793). Wiley. https://doi.org/10.1002/9781119482260

Makowski, D., Ben-Shachar, M. S., & Lüdecke, D. (2019). bayestestR: Describing effects and their uncertainty, existence and significance within the Bayesian framework. *Journal of Open Source Software, 4*(40), 1541. https://doi.org/10.21105/joss.01541

McCullagh, P., & Nelder, J. A. (1989). *Generalized linear models* (2nd ed.). Chapman & Hall. https://doi.org/10.1007/978-1-4899-3242-6

Metropolis, N., Rosenbluth, A. W., Rosenbluth, M. N., Teller, A. H., & Teller, E. (1953). Equation of state calculations by fast computing machines. *Journal of Chemical Physics, 21*(6), 1087–1092. https://doi.org/10.1063/1.1699114

Møller, J. (2013). *Spatial statistics and computational methods* (Vol. 173). Springer Science & Business Media.

Montgomery, D. C., Peck, E. A., & Vining, G. G. (2021). *Introduction to linear regression analysis*. Wiley.

Muth, C., Oravecz, Z., & Gabry, J. (2018). User-friendly Bayesian regression modeling: A tutorial with rstanarm and shinystan. *Quantitative Methods for Psychology, 14*(2), 99–119. https://doi.org/10.20982/tqmp.14.2.p099

O'Hara, R., & Kotze, J. (2010). Do not log-transform count data. *Nature Precedings*, 1. https://doi.org/10.1038/npre.2010.4136.1

Paananen, T., Piironen, J., Buerkner, P.-C., & Vehtari, A. (2020). Implicitly adaptive importance sampling. *Statistics and Computing, 31*, Article 16. https://doi.org/10.1007/s11222-020-09982-2

Peterson, M. (2017). *An introduction to decision theory*. Cambridge University Press. https://doi.org/10.1017/9781316585061

R Core Team. (2013). *R: A language and environment for statistical computing*. R Foundation for Statistical Computing. www.R-project.org/

Regazzini, E., & Bassetti, F. (2008). The centenary of Bruno de Finetti (1906–1985). *Scientifica Acta, 2*, 56–76.

ScienceBuddies. (2012). Probability and the birthday paradox. *Scientific American*. www.scientificamerican.com/article/bring-science-home-probability-birthday-paradox/#

Shiffrin, R. M., Lee, M. D., Kim, W., & Wagenmakers, E. J. (2008). A survey of model evaluation approaches with a tutorial on hierarchical Bayesian methods. *Cognitive Science, 32*(8), 1248–1284. https://doi.org/10.1080/03640210802414826

Slater, M., Rovira, A., Southern, R., Swapp, D., Zhang, J. J., Campbell, C., & Levine, M. (2013). Bystander responses to a violent incident in an immersive virtual environment. *PLOS ONE*, e52766. https://doi.org/10.1371/journal.pone.0052766

Sorensen, T., Hohenstein, S., & Vasishth, S. (2015). Bayesian linear mixed models using Stan: A tutorial for psychologists, linguists, and cognitive scientists. *Quantitative Methods for Psychology*, 175–200. https://doi.org/10.20982/tqmp.12.3.p175

Spizzirri, G., Eufrásio, R., Lima, M. C. P., de Carvalho Nunes, H. R., Kreukels, B. P., Steensma, T. D., & Abdo, C. H. N. (2021). Proportion of people identified as transgender and non-binary gender in Brazil. *Scientific Reports, 11*(1), 1–7. https://doi.org/10.1038/s41598-021-81411-4

Stan Development Team. (2011–2019). Stan modeling language users guide and reference manual 2.25. https://mc-stan.org

Stone, C. J. (1994). The use of polynomial splines and their tensor products in multivariate function estimation. *Annals of Statistics, 22*(1), 179–184. https://doi.org/10.1214/aos/1176325364

Van Ravenzwaaij, D., Cassey, P., & Brown, S. D. (2018). A simple introduction to Markov chain Monte–Carlo sampling. *Psychonomic Bulletin & Review, 25*(1), 143–154. https://doi.org/10.3758/s13423-016-1015-8

Vehtari, A., Gelman, A., & Gabry, J. (2017). Practical Bayesian model evaluation using leave-one-out cross-validation and WAIC. *Statistics and Computing, 27*(5), 1413–1432. https://doi.org/10.1007/s11222-016-9696-4

Wagenmakers, E.-J., Lodewyckx, T., Kuriyal, H., & Grasman, R. (2010). Bayesian hypothesis testing for psychologists: A tutorial on the Savage–Dickey method. *Cognitive Psychology, 60*(3), 158–189. https://doi.org/10.1016/j.cogpsych.2009.12.001

Watanabe, S. (2010). Asymptotic equivalence of Bayes cross validation and widely applicable information criterion in singular learning theory. *Journal of Machine Learning Research, 11*(December), 3571–3594.

Watanabe, T., Nanez, J. E., & Sasaki, Y. (2001). Perceptual learning without perception. *Nature, 413*, 844–848. https://doi.org/10.1038/35101601

Yildirim, I. (2012). *Bayesian inference: Metropolis-Hastings sampling*. Department of Brain and Cognitive Sciences, University of Rochester, Rochester, NY.

INDEX